Excursion Guide to the Geomorphology
of the Howgill Fells

Bowderdale Beck downstream of the Leath Gill confluence (see Chapter 6).

Snow covered 10th Century AD gullies at the head of West Grain, central Howgills (see Chapters 2, 6)

CONTENTS

LIST OF TABLES

ABOUT THE AUTHOR

Professor Adrian Harvey graduated from University College, London in 1962 with a BSc Geography (with Geology) degree, and in 1967 with a PhD in Geomorphology and Hydrology. From 1965 he worked in the Department of Geography (latterly the School of Environmental Science) at the University of Liverpool, until his retirement in 2005 as Professor of Geomorphology. His research career focuses on modern geomorphic processes and recent landforms. In addition to much research in dry regions, particularly in Spain and parts of the USA, a main research emphasis within the UK has been on relations between hillslope gully erosion and stream and valley system response within the Howgill Fells of Cumbria. That research had two threads: (i) the modern process system, based on sustained monitoring, and (ii) together with colleagues, the study of landform change and evolution over the last few thousand years or so, in response to human- and climatically-induced environmental change. Professor Harvey has over 100 research publications, more than 20 of which relate to the Howgill studies.

ACKNOWLEDGEMENTS

In preparing this book, I have many people to thank for their assistance and collaboration over many years. First, I have to thank the many academic colleagues with whom I have worked in the Howgills and other areas in the north of England over the years, both from the University of Liverpool (Frank Oldfield, Peter James, Nigel Richardson, Peter Cundill, Richard Chiverrell, who has now taken over my previous role co-ordinating Howgill research), and from elsewhere (Roy Alexander, Steve Wells, Bill Renwick), plus former research students from the University of Liverpool (including Asma al Farraj, Suzanne Hunter née Miller, Gez Foster, Jennie Millington). I also profoundly thank Sandra Mather, who prepared the illustrations for this book and for my previous publications – without her help, I would be lost! My thanks also go to Anthony Kinahan of Dunedin Academic Press, for thoughtful editorial work. Finally, my deep-felt thanks go to my wife, Karina, for her patience and tolerance not just in relation to this book, but to years of research-related demands!

Part One
Geomorphology of the Howgill Fells

CHAPTER 1

Introduction – regional overview

Location

The Howgill Fells (Frontispiece, Figs 1.1, 1.2) form a spectacular triangular block of hill country in eastern Cumbria in northern England. There are no roads within the Howgills, so few people (serious hikers and geomorphologists excepted) visit the Howgills. However, many have seen their dramatic western edge, which forms the impressive scenery of the Lune gorge, through which pass both the M6 motorway and the West Coast main railway line (Fig. 1.3). The Howgills lie between the Lake District and the Yorkshire Dales. The southern part of the Howgills had long been part of the Yorkshire Dales National Park, but recently National Park status has been extended to include the whole of the Howgills. Over a period of more than 30 years during my career at the University of Liverpool, together with colleagues from both Liverpool and elsewhere, I studied the geomorphology of the Howgills. Partly based on the results of those studies, a large area within the north-western Howgills was recently designated as an SSSI (Site of Special Scientific Interest). The available published topographic and geological map cover for the Howgills is given in Appendix 1.

Bedrock Geology

Geologically, the Howgills form an eastward extension of the Lower Palaeozoic rocks and the structures of the Lake District (Fig. 1.4) (Edwards and Trotter, 1954). These rocks and structures are bounded to the east by a major regional structure, the Dent fault, beyond which lies the Carboniferous Limestone country of the core area of the Yorkshire Dales National Park. The bedrock geology is not particularly striking. Except in the Cautley area in the east of the Howgills, there are few good outcrop exposures. The rocks themselves are mostly a rather monotonous sequence of folded Silurian siltstones with some grits, sandstones, shales and an occasional thin muddy limestone horizon (Table 1.1). The uppermost of the Silurian rocks, outcropping along the northern Howgill margin, are slates. The structure is broadly anticlinal along roughly an E–W axis, but it is complicated by numerous Caledonian minor folds orientated NE–SW. Several of these folds in the Cautley area expose the underlying Upper Ordovician (Ashgillian) shales and thin limestones, plus thin volcanic rocks, primarily of acid to intermediate composition. To the north of the Howgills, the folded Silurian rocks are unconformably overlain by Carboniferous Limestone. These rocks dip gently north, forming a pronounced escarpment bounding the upper Lune valley between Ravenstonedale and Orton, north of Tebay (Figs 1.1, 1.5). Along the Rawthey valley in the

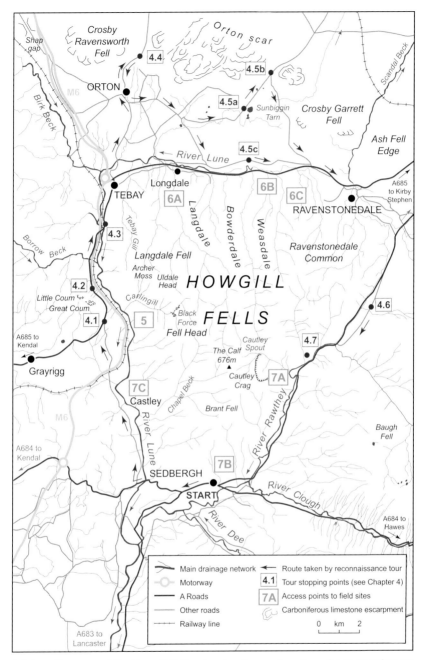

1.1 Location map of the Howgill Fells, showing the road system, the route taken on the reconnaissance field excursion (Chapter 5) and the access points for the detailed field excursions (Chapters 5–7) (numbered by chapter and stop number).

Sedbergh area a basal Carboniferous conglomerate also rests unconformably on the folded Silurian rocks.

The bedrock geology, though interesting, is not the reason for this book. It is through their geomorphology that the Howgills stand out as a superb

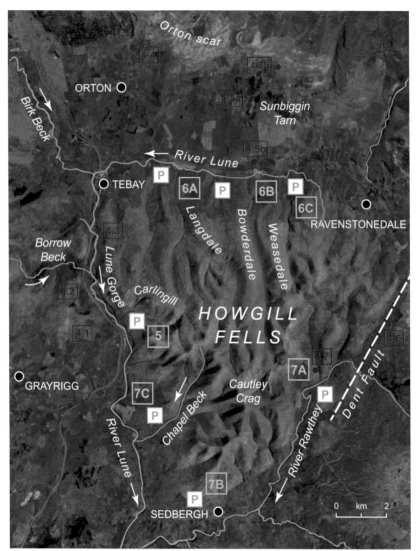

1.2 Google Earth image of the Howgill Fells and adjacent areas, showing the general locations of the field excursions.

1.3 The Lune Gorge, looking north towards Tebay: the Howgill Fells are on the right (east). Through the gorge on the skyline is the Carboniferous Limestone escarpment of Orton Scar.

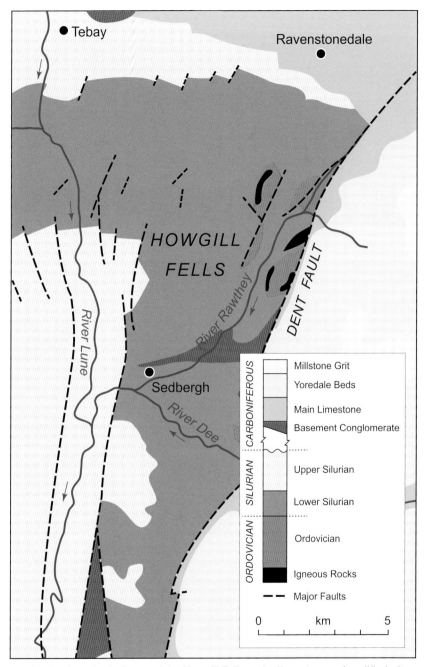

1.4 Bedrock geological map of the Howgill Fells and adjacent areas (modified after Edwards and Trotter, 1954).

field site, particularly in the context of modern processes and of recent landform evolution. However, before we can consider modern and recent geomorphic processes we need to consider some of the broader aspects of the landforms.

1.5 Looking south from Orton Scar, down the scarp face of the escarpment of the Carboniferous Limestone, across the upper Lune Valley (the River Lune here flows east to west/left to right). Looking towards the Howgill Fells beyond. The skyline is the summit erosion surface of the Howgills.

Glaciation

During the Pleistocene, as everywhere else in northern Britain, the Howgills were glaciated a number of times (King, 1976). We know very little about the earlier glaciations in this area, but during the peak of the last glaciation (between c.26 and 17ka BP) the Howgills and the whole of the surrounding area were covered by an ice sheet that extended from Scotland and the Lake District, SE towards York, and south into the west Midlands (Chiverrell and Thomas, 2010). At that time, of course, global sea levels were low, and Britain was merely a peninsula of northwest Europe.

There has been some recent research on the dynamics of the ice sheet affecting Cumbria during the last glaciation (between c.26 and 17ka BP). Over that period of time the effectiveness of the different ice sources varied, causing variations in the strength of the several ice streams involved, and therefore in the resulting patterns of basal ice erosion and deposition. North of the Howgills, the ice sources involved were the Lake District itself and the Southern Uplands of Scotland (Evans *et al.*, 2009; Livingstone *et al.*, 2012). The interactions between these two ice sheets affected the ice streams that entered the Eden valley and which ultimately passed east over the Pennines through the Tyne gap or over the Stainmoor pass. During the final phases the Lake District sources weakened, resulting in the so-called Scottish re-advance into north-eastern Cumbria (Livingstone *et al.*, 2010). The Howgills were probably an ice divide, although south of the Howgills there was interaction between Lake District ice, Pennine Ice and further south still, the so-called Irish Sea ice sheet (Chiverrell *et al.*, 2016).

The direct effects of glaciation include intense glacial erosion in the source areas (e.g. in the Lake District), but with little obvious evidence of glacial erosion in the Howgills, except for the cirque at Cautley Crags (Fig. 1.6) and perhaps several rather crude cirque-like forms on the west

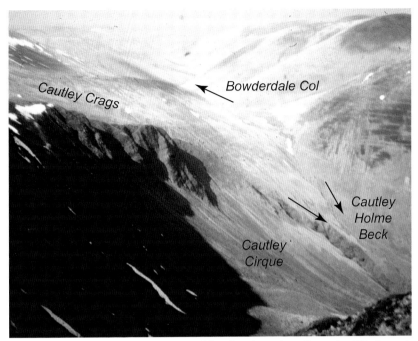

1.6 Looking north from the south side of Cautley cirque (see Excursion 7A), through the col at the head of Bowderdale. The col was abandoned after glacial diversion of the Bowderdale headwater drainage into the Cautley valley.

flank of the Lune gorge, southwest of Tebay. The depositional effects within the Howgills themselves were limited. Glacial deposition of boulder clay did occur along the major valleys. Especially within the larger valleys and lowland areas outside the Howgills, the deposits were locally often moulded into whaleback drumlin forms. Such forms are common in the Eden Valley to the north of the Howgills, also in the Kendal area west of the Howgills, and along the Lune valley near to and south of Sedbergh. Perhaps the most significant 'glacial' alterations to the landscape in our area were the erosional effects of glacial meltwater, acting perhaps sub-glacially. There are a number of examples in our area of diverted drainage, involving diversions from north-flowing to south-flowing drainage, which may have had their origins in meltwater-induced erosion (see below).

The ice sheet eventually died back. We have some indication of the time from which the Howgill area had become ice-free. Recent work by Wilson *et al.* (2013), using a relatively new technique of 'cosmogenic dating', whereby the presence and abundance of cosmogenically derived radioactive isotopes (particularly Berryllium 10: ^{10}Be) on exposed rock surfaces can give an estimate of the duration of exposure. They applied this technique to Shap Granite erratic boulders perched on the Carboniferous Limestone escarpment north of Orton, and derived an estimated age of deglaciation of *c.*17ka BP.

Following deglaciation, cold climates intermittently persisted, culminating in the final regional glacial phase, the so-called Loch Lomond re-advance (*c*.11ka BP) at the end of the Pleistocene. At that time the Lake District cirques were re-occupied by ice (Sissons, 1980). Apart from the possibility of a small cirque glacier at Cautley (Fig. 1.6) there is no evidence for 'Loch Lomond' (11–10ka BP) stage ice in the Howgills. Over the late-glacial period too, sea levels were rising and continued to rise into the early Holocene, ultimately forming Morecambe Bay and the Solway Firth, into which the Howgill river systems drained.

With continuing cold climates and little or no actual glaciation, as elsewhere in northern England (Harrison, 1996) intense periglacial processes dominated the Late Pleistocene. On the limited bedrock outcrops freeze-thaw weathering produced scree debris, accumulating on the slopes below. Much more widespread were the effects of permafrost: freezing and thawing of the active layer, especially on valley-side boulder-clay deposits, generated a mobile surface layer that sludged downslope by seasonal 'solifluction'. Virtually all the valley sides within the Howgills are blanketed by a mantle of solifluction material. The valley sides are characterized by smooth slopes over a semi-stratified stony clay deposit (see below, Chapter 2 and Figs 1.7, 1.8a,b). Another feature related to permafrost is patterned ground. This is characteristic of relatively stony material on low-angle slopes. On refreezing, the moisture content of the matrix forces the stones to move laterally until the stony areas and the matrix-rich areas are separate. On gentle slopes this process forms stone polygons, with stone garlands on slightly steeper slopes. Such features are well developed on the flanks of Bowderdale (see Fig. 1.8c; see also Chapter 6), below scree-mantled slopes. At the same time during the late Pleistocene, the streams were depositing sorted gravels on the then valley floors. These have since been dissected to form the highest of the stream terraces, of which patches now remain.

1.7 Lower Bowderdale looking north. The gently sloping surface to the west (left) of the stream is the Late Pleistocene solifluction surface, into which the modern valley is cut. There is a low (Holocene) stream terrace on the right. Note the stream channel: a stable single-thread channel (see Chapter 3).

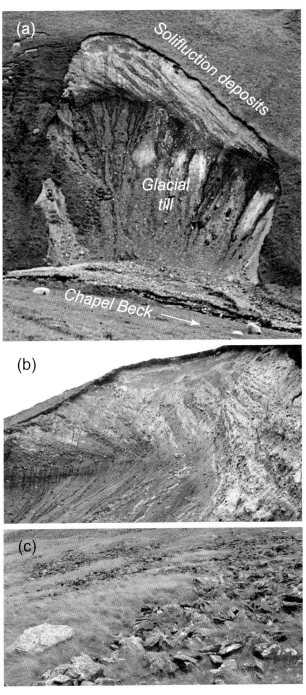

1.8 Periglacial effects. a) Erosional section cut by Chapel Beck (which flows west/to the right), exposing glacial till at the base overlain by dipping solifluction sediments; b) Detail: the solifluction sediments, Chapel Beck (see Chapter 7C); c) Periglacial patterned ground – stone garlands above Bowderdale (see Chapter 6B).

At the end of the Pleistocene (10ka BP) climate ameliorated into the sequence of temperate climatic fluctuations of the Holocene. The solifluction slopes stabilized and became vegetated, setting the scene for the Holocene sequence I deal with later (see Chapter 2).

Pre-glacial topography

Having considered the effects of glaciation, one important question that remains is 'what kind of landscape preceded Pleistocene glaciation?' This topic in general was the focus of much geomorphological research in Britain up to about the 1960s. It had long been recognized that throughout Britain, especially in the upland areas, the summit areas of the hills are generally relatively low-slope plateau areas. Sometimes these plateau areas coincide with more or less horizontal underlying bedrock, but generally the gently sloping surfaces truncate the underlying structures, indicating that they are erosional features (erosion surfaces), presumably related to some ancient geomorphological regime. Until around the 1960s there was much debate in the geomorphological world about their possible origins. One possibility is that perhaps they may represent uplifted former marine wave-cut surfaces, either extant, or after erosional removal of a weak sedimentary cover. Another possibility is that they are perhaps sub-aerially eroded pediment (or peneplain) surfaces related to former fluvial base levels, formed during the Tertiary, when base levels were more stable than during the Pleistocene. A third possibility is that perhaps they are the result of the deep weathering of the underlying bedrock during the tropical/sub-tropical climates of the mid-Tertiary (etchplains), with later erosional removal of the weathering mantle. The origin of the erosion surfaces is still far from clear.

In the Howgills, a series of erosion surfaces has long been recognized that bevel the summit areas and the northern ridges (Fig. 1.9) (McConnel, 1939), cross-cutting the underlying structures in the Silurian rocks. When previous topography is reconstructed (King, 1976) these surfaces appear to relate to drainage from a major E–W watershed across the centre of the Howgills, northwards towards the Eden valley and southwards towards the Lune. As elsewhere, the origins and age (presumably Mid-Late Tertiary?) of these surfaces are far from clear.

1.9 The Howgill Fells from the west, looking across the Lune valley into the Chapel Beck valley system (see Chapter 7) (the Lune flows south – left to right). Note the skyline: the pre-Pleistocene Howgill summit surface.

Drainage evolution

The present drainage pattern of the Howgills is dominated by the south-flowing River Lune and its tributaries, especially the Rawthey on the eastern side of the fells (Fig. 1.10). Only a tiny part of the north-eastern Howgills

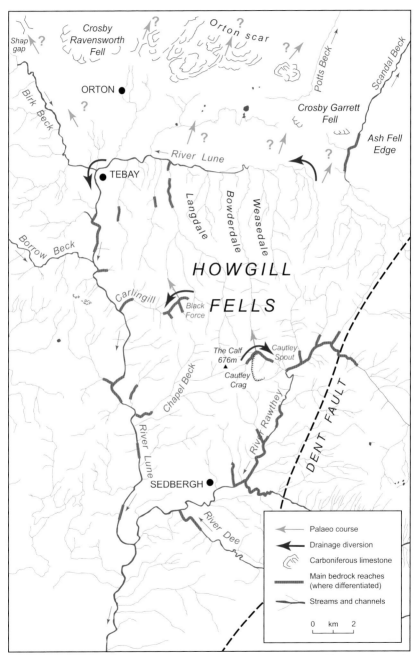

1.10 Map of the modern drainage pattern in the Howgill Fells, picking out the main zones of channels cut into bedrock and the postulated Pleistocene changes in drainage directions.

near Ravenstonedale currently drains north through Scandal Beck to the Eden drainage. Dominance by south-flowing drainage has not always been the case. During the Late Tertiary, if one believes the erosion surface story, the northern Howgills drained north towards the Eden. The upper Lune subsequent valley along the base of the Carboniferous rocks did not exist. The main headstream of the Lune itself presumably was the Borrow Beck drainage, from the northwest. The conventional wisdom (King, 1976) is that, probably during the early Pleistocene, the upper Lune developed as a subsequent stream along the basal Carboniferous unconformity to the north of the Howgills. It successively captured the northern Howgill tributaries, west to east. These streams had previously flowed through gaps in the limestone escarpment on their way towards the Eden. I wonder though, whether this interpretation is correct – there is no direct evidence for it. In my opinion, equally feasible would be for the subsequent stream to have developed earlier as the Lowther headwater, cutting back through the Shap gap, along the base of the Carboniferous outcrop to capture the northern Howgill streams. Under that scenario, the present course of the Lune through the Lune gorge south of Tebay would have been a glacial diversion, perhaps from an early glaciation. There have been two other captures/diversions, though at a much smaller scale, both of them similarly resulting in the dominance of south-flowing drainage. One of these is the capture/diversion of the head of Uldale at Blakethwaite into Carlingill (see Chapters 2, 4); the other is the glacially induced capture/diversion at the head of Bowderdale Beck by the Cautley/Rawthey drainage (see Chapters 2, 6 and 7 and Figs 1.6, 1.10).

I deal with the evolution of the modern landscape, since the Late Pleistocene glacial conditions, in the next chapter.

Table 1.1: Bedrock geology of the Howgill Fells and adjacent areas (after Edwards and Trotter, 1954).

Carboniferous	Millstone Grit (summit area of Baugh Fell, E of Dent fault)
	Yoredale Series (E of Dent fault)
	Main limestone (N of Howgills and E of Dent fault)
	Basal conglomerate (Sedbergh area and Tebay area)
	Unconformity
Silurian	Kirkby Moor Flags (outcrop only to the west of the Howgills)
	Bannisdale slates (Lune valley and N Howgills)
	Coniston flags, Coniston grits (Howgills)
	Stockdale shales (NE Howgills)
Ordovician	Ashgillian shales (NE Howgills)
Igneous Rocks	(age uncertain – Ordovician/Silurian?)
	Intrusive felsites, etc.

CHAPTER 2

Holocene landform evolution

In this chapter I deal with landform evolution within the Howgills, from the periglacial period at the end of the Pleistocene (see previous chapter), up to the present day, but excluding modern processes (to be dealt with in Chapter 3). I deal with two main phases of geomorphic activity: first with landform development from the late Pleistocene to the early Holocene, then with the later Holocene sequence in relation to contemporaneous environmental change. However, before considering the specifics of the Howgill geomorphology, we need to consider the nature of the evidence used in general to establish such landform developmental sequences.

Evidence for the sequence of landform evolution

The evidence used to establish any sequence of landform development is threefold: (a) the landforms themselves, including their relative positions and composition; (b) some form of relative dating of the surfaces or the constituent sediments to establish temporal correlations amongst the landforms; (c) some form of absolute dating, usually of the constituent sediments to establish at least parts of a real chronology.

Below I consider each of these types of evidence first in general terms, then in relation specifically to the post-glacial Howgill story.

(a) Evidence provided by the nature of the landforms themselves

Several questions can be asked about the landforms. Are they erosional or depositional landforms? Are they individual forms, or can they be related to similar forms, in similar positions? If erosional, do they truncate recognizable older landforms and sediments? If depositional, are they cut into older forms or do they simply bury them? What is their sedimentary composition? What can we infer about the mechanism and environment of deposition? For both erosional and depositional landform suites it is important to establish the answers to these questions, in order later to make inferences about the chronology of landform development and to assess any environmental implications. Particularly important for depositional landforms are the sedimentary properties. What are the particle sizes present? If there are clasts present, are they of local or 'foreign' bedrock geology? What is the relationship between clasts and matrix (clast- or matrix-supported)? What is the internal organization of the sediments (bedded or massive – dip of beds in relation to topographic form)? What is the clast fabric (i.e., attitude of clasts, relationships to one another and to bedding)? Answers to these questions will help to clarify the sediment transport and depositional mechanism. Within the Howgills, there are suites of superbly developed landforms, both erosional (e.g. gully systems) and depositional (e.g. alluvial fans and stream terraces) that post-date the latest glacial ice cover, and represent the late Pleistocene and Holocene geomorphic sequence. They form the basis of this field guide.

(b) Aids to relative dating of the landforms

There are several types of evidence that can provide information on the relative age of the landforms, and on the environmental sequence within which the landforms developed. Are there properties of the landforms themselves that enable an estimate of their relative ages to be made, at least in relation to other similar landforms? Is there evidence that can yield information of the environmental sequence contemporaneous with the landform developmental sequence? The first two aids presented below (soil development and lichenometry) relate to the landforms themselves; the third aid (the vegetation sequence) relates to the contemporary environmental sequence.

Soil development. Once erosion or deposition ceases and the surface becomes relatively stable, soil formation begins. In general of course, the rate of soil development depends on a whole range of factors such as parent material, climate, etc. However, within a particular environment these factors may be similar on different local landforms. Within the Howgills on stable reasonably drained sites the zonal soil is a podzol (Miller, 1991), an acidic soil that shows clear horizon development

2.1 Mature podzol profile developed on the Late Pleistocene High Terrace of Carlingill.

(Fig. 2.1). In the Howgills well-developed mature podzols are restricted to the surfaces of Late Pleistocene landforms. On successively younger Holocene terrace surfaces a chronosequence of immature podzolic soils can be identified (Harvey *et al.*, 1984; see also Fig. 2.2).

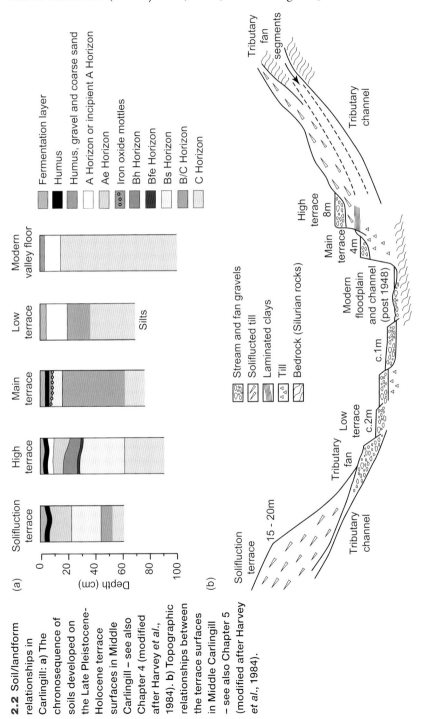

2.2 Soil/landform relationships in Carlingill: **a)** The chronosequence of soils developed on the Late Pleistocene-Holocene terrace surfaces in Middle Carlingill – see also Chapter 4 (modified after Harvey et al., 1984). **b)** Topographic relationships between the terrace surfaces in Middle Carlingill – see also Chapter 5 (modified after Harvey et al., 1984).

Lichen colonization. A useful age indicator on stony surfaces is lichen development, particularly the diameter of the thallus of the lichen *Rhyzocarpon geographicum* (Fig. 2.3 [inset]), a lichen that colonizes bare rock or stony surfaces. What is more, not only can lichenometry be used to differentiate surfaces on the basis of relative age, but an estimate of absolute age may be achieved by calibration of the growth rates from historical data, such as can be derived from dated gravestones (Fig. 2.3). In the Howgills, we have used lichenometry to differentiate younger Holocene (historical) stony surfaces, especially those on abandoned gravel bars on the valley floors (see Harvey *et al.*, 1984). This has allowed former stream channel patterns to be identified and dated.

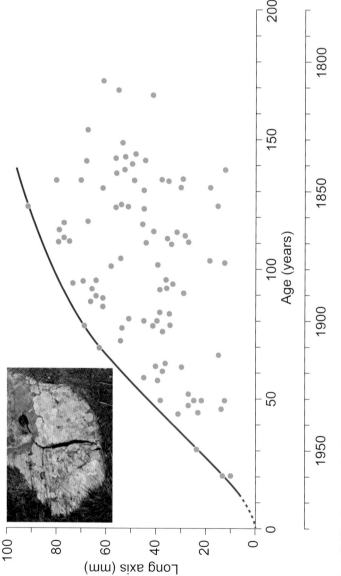

2.3 Calibration curve for the lichen *Rhyzocarpon geographicum*, derived from local graveyard data (Harvey *et al.*, 1984). Inset shows mature *Rhyzocarpon geographicum* lichens encrusting a boulder.

Pollen analysis. This is a technique that can be used on peaty sites for the reconstruction of the ambient vegetation sequence, from the pollen grains preserved within peats and peaty soils. A sample of A sample of the peat sequence or the peaty soil is taken in the field, usually in the form of a monolith block or a core. It is carefully labelled and taken back to the laboratory. There, it is subsampled and each subsample is treated to separate pollen grains from the peaty mass. Then for each subsample the pollen grains present are identified under the microscope and counted. Usually then, the data from each subsample are put together so that progressive changes in the presence or the abundance of pollen species can be identified, in turn highlighting any progressive change in the ambient vegetation cover over the timescale of the peat or peaty soil formation. This is not a relative dating technique in itself, but does enable reconstruction of past vegetation sequences, from which climatic and land-use inferences can be made. A pollen sequence has been derived from Archer Moss in the western Howgills covering the last almost 4000 years (Cundill, 1976; Chiverrell *et al.*, 2007, 2008). Other sites too have yielded similar information of the vegetation sequence (Chiverrell *et al.*, 2007, 2008). All have been extremely useful in reconstructing the late Holocene and Historical environmental sequence for the Howgills (see later, this Chapter and Chapters 5 and 6).

(c) Methods of absolute dating

The conventional method of absolute dating of Late Quaternary sediments is radiocarbon dating of organic matter. This analysis depends on the existence of two isotopes of carbon (^{14}C and ^{12}C), which in living matter maintain a constant ratio, but on death the radioactive ^{14}C decays to ^{12}C at a known rate, thus allowing the time since death to be established. The ratio between ^{14}C and ^{12}C gives an estimate of the time since death of the organism, expressed in Radiocarbon Years before present (BP: prior to 1950). However, partly because of variations of atmospheric CO_2 in the past, Radiocarbon Years BP have to be calibrated to calendar years BP, using one of several calibration methods (e.g. Stuiver and Reimer, 1993; Stuiver *et al.*, 2005). For the dates we have in relation to sites within the Howgill Fells, in this book I quote them in Radiocarbon Years, but in a summary table (see Table 2.1) I give calendar dates too.

Several horizons from the peat, from which the pollen sequence at Archer Moss was derived, have been radiocarbon dated. The dating has allowed the Holocene vegetation and environmental chronology to be firmly established. Perhaps more important from the geomorphic point of view, we have had a number of radiocarbon dates from organic horizons of buried soils interstratified within the sediments comprising the Holocene landform sequence. These have been especially from sites in the western and northern Howgills. They have allowed a broad chronology of the geomorphic sequence to be established (see Table 2.1; see also Harvey *et al.*, 1981; Harvey, 1996; Chiverrell *et al.*, 2007, 2008).

Late Pleistocene to Early Holocene landform evolution
(periglacial processes then hillslope stabilization)

The previous chapter established that the end of the last regional glaciation occurred around 17ka BP. Within the Howgills, except at Cautley (see Chapter 7), there is little direct evidence of glacial erosional features, nor of specific depositional morainic features. However, deposition of glacial till (boulder clay) at the base of the mobile ice sheet was important along most of the major valleys. Meltwater erosional activity was also probably important in modifying the drainage (see Chapter 1).

The only real evidence of a local ice source is the cirque and the moraines at Cautley (see Fig. 1.6; see also Chapters 4 and 7). Whether these features relate to the persistence of a local glacier at the close of the regional glaciation (the most probable situation) or to a later small glacier at the very end of the Pleistocene during the so-called Loch Lomond stage, when there were corrie glaciers within the Lake District, is uncertain. There are also a couple of cirque-like forms on the western slopes above the Lune Gorge south of Tebay, which may have supported small short-lived local glaciers at the close of the regional glaciation (see Chapter 4).

For much of the period from the decay of the regional ice cover (c.17ka BP) until the end of the Pleistocene at about 10ka BP, and especially during the intense cold of the Loch Lomond stage (c.11–10ka BP), there would have been permafrost (see Chapter 1). The permafrost would have been subject to seasonal thawing of the active layer near the surface, resulting in slope instability and the seasonal movement downslope of the wet, unconsolidated, ill-sorted, reworked glacial till, by solifluction. The results of this process were the crude into-valley stratification of the solifluction sediments, and the formation of sloping into-valley surfaces (see Chapter 1; Figs 1.7; 1.8a,b). Another process related to permafrost is the formation of patterned ground (also described in Chapter 1; see Fig. 1.8c). Such features are well developed on the flanks of Bowderdale (see Fig. 1.8c; see also Chapter 6), below scree-mantled slopes. At the same time the streams were depositing sorted gravels on the then valley floors, forming the sediments of the modern high terrace (see Chapter 1).

With the end-Pleistocene climatic amelioration and the thawing of the permafrost, solifluction ceased and the hillslopes stabilized. Vegetation colonization took place. Both processes greatly reduced sediment supply downslope and to the stream channels. With reduced sediment load the streams incised into their former beds, leaving their former sediments perched as the high terrace. As solifluction surfaces and the high stream terrace ceased to undergo deposition at the end of the Pleistocene, both now support relatively mature podzol soils (see Fig. 2.1) that have developed throughout the Holocene. At the same time there were vegetation changes, adding to hillslope stabilization (see later section in this chapter, on the Holocene vegetation sequence).

The later Holocene
(including several waves of hillslope gullying)

Much later in the Holocene there were waves of hillslope gullying, including a major wave of gullying during approximately the tenth century AD, and probably several minor waves. The resulting gully features are exceptionally well developed in the Howgills (Figs 2.4, 2.5). Similar features do occur in other parts of northern Britain, for example in the Forest of Bowland on the Lancashire/Yorkshire border and in the Southern Uplands of Scotland (Chiverrell *et al.*, 2007). Nowhere south of the Scottish Highlands are they as well developed as they are in the Howgills. The gully systems cut into the Late Pleistocene soliflucted till on the hillslopes and range in form from simple valley-side linear

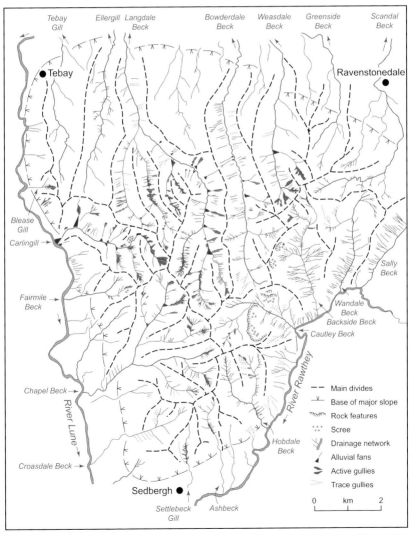

2.4 Distribution of now-stabilized gully systems in the Howgill Fells (modified after Harvey, 1996).

2.5 Morphology of the older gully systems: the example of Great Ulgill, head of Carlingill (see Chapter 5); note the contrasts between valley-side and valley-head networks.

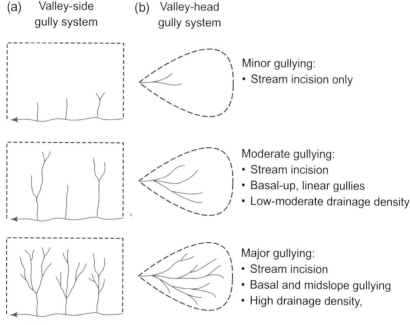

(a) Valley-side gully system

(b) Valley-head gully system

Minor gullying:
• Stream incision only

Moderate gullying:
• Stream incision
• Basal-up, linear gullies
• Low-moderate drainage density

Major gullying:
• Stream incision
• Basal and midslope gullying
• High drainage density,

2.6 Styles of gully network patterns: for low (minor or no gullying), moderate gullying, and major gullying of high drainage density (modified from Harvey, 1996) for: a) valley-side gully systems; b) valley-head gully systems.

gullies to complex valley-head dendritic networks (Figs 2.5, 2.6). Some of the smaller gully systems appear to be basally controlled, but the majority appear to have developed in mid-slope. Network complexity appears to be a reflection of slope length (valley-side gullies) or drainage area (valley-head gullies) and slope steepness. In an attempt to identify how these factors influenced gully network development, the valley-side and valley-head networks were classified on the basis of their complexity (Harvey, 1996) (Fig. 2.6). Then their relations to the key variables were considered (Fig. 2.7 a,b). For the valley-side gullies it is clear that slope gradient is the critical variable. Approximate threshold

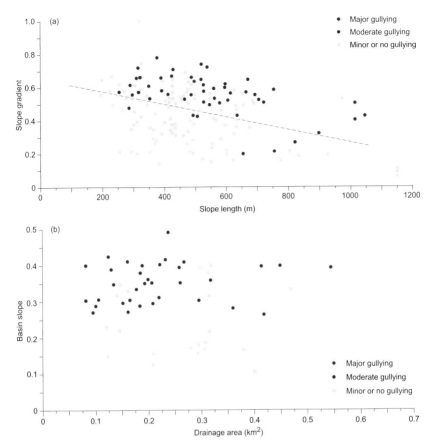

2.7 Influence of slope length (valley-side gully systems), drainage area (valley-head gully systems) and slope/gradient (modified from Harvey, 1996) for: **a)** valley-side gully systems; **b)** valley-head gully systems.

conditions for the formation of complex valley-side gully networks can be identified whereby the critical slope is slightly less for greater slope lengths. The approximate threshold can be described by the simple equation:

$$S = 0.62 - 0.31L$$

[where S is slope gradient (ratio), and L is slope length (km) (Fig. 2.7a)].

For the valley-head situation a similar approach was used (Fig. 2.7b), but rather than considering actual slope gradients the mean basin slope has to be used as a variable. There does seem to be an approximate minimum threshold value, but it is much less clearly defined than in the case of valley-side gullies. Drainage area seems to have little influence. The actual threshold mean basin slope for valley-head gully systems seems to be considerably less than the threshold slope gradient for the valley-side case. That might simply be an artefact of the different geometries between the two situations or a function of convergent runoff within the valley-head case. In both cases the dominance of the slope variable as a control suggests that on-slope runoff power was the dominant control, rather than basal coupling (see Chapter 3). Increased runoff is a situation that could have been produced as the result of dramatic vegetation change (see Holocene vegetation, below) induced, for example, perhaps by climate change or perhaps by overgrazing?

Hillslope gullies are particularly well developed within the Carlingill system (see Chapter 5) and the northern valleys (see Chapter 6). Since their formation they have now largely become revegetated and stabilized. The former gully systems fed sediment into the stream system to cause aggradation on the valley floors, as gravel sediments which have since been dissected to form the lower terraces, evident particularly in Carlingill (see Chapter 5), Langdale and Bowderdale valleys (see Chapter 6). Soils on these terraces are much less well developed than the mature podzols on the high terrace (Fig. 2.2; see also above and Miller, 1991). In addition, where large gully systems fed excessive sediment into the valley system, not all the sediment was incorporated into the stream sediment load, but instead accumulated in alluvial fans at the foot of the hillslope gullies. Large fans were formed in Langdale and Bowderdale (Fig. 2.8; see also Chapter 6), in

2.8 Burnt Gill Fan, Langdale: a) general view; b) detail of buried soil: the soil formed on stream gravels of the low terrace (base of photo); the soil (dark layer) is overlain by alluvial fan gravels (see also Fig. 6.2).

2.9 Late Pleistocene–Holocene landform relationships in Middle Carlingill, including the Late Pleistocene high terrace, cut and inset by Haskaw Gill tributary fan. Carlingill, at the base, flows to the west (to the right) (see also Fig. 2.2b and Chapter 5).

Carlingill at the Grains Gill confluence (see Frontispiece, see also Chapter 5) and in Middle Carlingill (Fig. 2.9), also at Blakethwaite at the head of the Carlingill valley system (Fig. 2.10; see also Chapter 5). Organic layers from within the soils buried by these fan deposits have been radiocarbon dated. At Grains Gill, fan deposition took place after 2145 +/-40 BP. At a section at Burnt Gill in Middle Langdale (Fig 2.8; see also Chapter 5) the older of the two dates, from the base of the soil, is 2580 +/-55 BP, and the youngest date from the top of the soil, immediately prior to burial by the fan deposits, is 940 +/-95 BP (Harvey et al., 1981). A further date of 1042 +/-22 BP from the top of a buried soil at an adjacent site (Chiverrell et al., 2007) confirms this overall sequence. These dates indicate that the Burnt Gill site was stable for at least 1500 years prior to burial, but that gully erosion on the hillslope above fed sediment to the fans in about the tenth century AD. At Blakethwaite we have a series of dates from the buried soil sequence. The oldest, from the base of the soil, is 2975 +/-85 BP. The main date from the humic fraction at the top of the underlying buried soil is 1210 +/-50 BP (Harvey, 1996), but a date from the humin fraction of overlying buried soil of 890 +/-50 BP. This indicates burial during the intervening period. There is a date of 410 +/-50 BP from the humic fraction of the same soil, immediately prior to burial by a secondary minor set of gravels. This suggests a minor pulse of sediment addition following that date (see below). Furthermore, we now have 5 dated sites from within the Chapel Beck catchment in the SW Howgills (Chiverrell et al., 2008). In total within the Howgills, there are now 22 dates available, from 17 buried soils or other

2.10 Blakethwaite alluvial fan: **a)** general view; **b)** detail of section with soil buried by fan deposits (see also Chapter 5).

organic materials at 11 sites, spanning an age range of landforms from the high terrace to the modern valley floors (Chiverrell *et al.*, 2008). What is significant is that for the 9 dated soils buried by debris cones, 6 (including the main Blakethwaite and Langdale sites) are indications of what appear to be several phases of erosion. Particularly significant is a major phase of hillslope gully erosion and fan deposition initiated during the tenth century AD, following a period of stability of at least 1000 years. The sequence is summarized in Table 2.1.

At some time after the tenth century AD, the system largely stabilized again, most of the gullies presumably became vegetated and ceased to supply sediment to the fans or the stream system. With the fan surfaces stabilized, the streams incised into the valley floors, leaving the former valley floor as a low terrace. Locally there seems to have been minor reactivation of some of the gullies, feeding relatively small amounts of sediment to the alluvial fans. At Blakethwaite at 410 BP (Harvey, 1996) and at two sites in the Chapel Beck valley at 275 and 280 BP (Chiverrell *et al.*, 2008) there were minor additions of sediment to the fans (see Table 2.1).

Much more recently, new hillslope gullies developed, though of much more limited extent than the tenth-century systems (see Chapter 3), and basally induced rather than induced in mid-slope. These now supply sediment to the modern fluvial system. I will deal with modern processes in the next chapter, but before that we need to consider the vegetational context for the late Holocene geomorphic sequence.

The Holocene vegetation sequence [Written in conjunction with Richard Chiverrell˙]

We have no direct evidence for the nature of the early–mid-Holocene vegetation in the Howgills, but there is evidence from neighbouring regions (e.g. by Turner and Hodgson, 1979, 1983, 1991, from the north Pennines; by Pennington, 1991, from the English Lake District; and by Oldfield, 1963 from north Lancashire). The pattern from these nearby regions shows that at the Pleistocene–Holocene transition (*c.*11ka BP), with the thawing of the permafrost and the cessation of hillslope solifluction, the Howgills would have experienced woodland colonization. This would have led eventually by the mid-Holocene to at least a partially wooded landscape dominated by hazel, birch, oak, elm and alder.

From the mid- to late Holocene we have much more information. There are six sites within the Howgills for which pollen sequence diagrams are available, including those from Archer Moss (Cundill, 1976; Chiverrell *et al.*, 2008) and Bowderdale (Miller, 1991), for which relatively long records are available. The other sites relate to shorter sequences. The oldest local information comes from a peaty profile developed on top of the (latest Pleistocene?) high fluvial terrace in Bowderdale (Miller, 1991). Several radiocarbon ages are available from this profile, from an early to mid-Holocene date of 6070 +/-70 BP, with intermediate dates of 3320 +/-70 BP and 2205 +/-40 BP and the youngest of 395 +/-50 BP. For the lower part of the profile, tree pollen (dominantly: *Alnus*/alder, *Betula*/birch, *Corylus*/hazel and *Quercus*/oak) accounts for almost 40% of the pollen rain, indicating a mostly wooded landscape. Following the middle dates there was a marked decrease in the tree pollen (to less than 20%), and an increase particularly in *Ericacea*/heathers.

A complete pollen profile for the mid-late Holocene is available from Archer Moss, on the Howgill summit surface north of Carlingill (Cundill, 1976; Chiverrell *et al.*, 2008) (Fig. 2.11). Above a basal date of 3480 BP several other dates are now available from this profile (Chiverrell *et al.*,

* Professor Richard Chiverrell: Dept of Geography and Planning, School of Environmental Sciences, University of Liverpool.

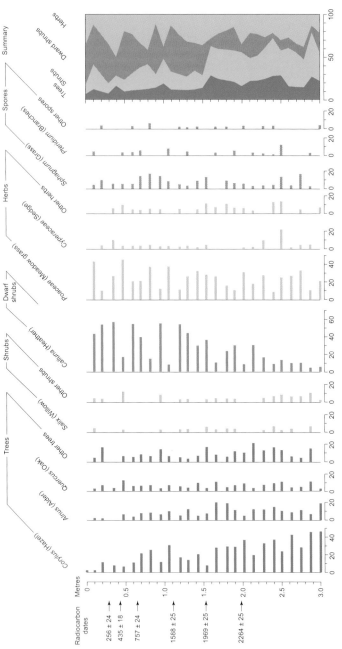

2.11 Archer Moss pollen diagram (modified from Chiverrell *et al.*, 2008, adapted from Cundill, 1976). Radiocarbon-dated (BP) horizons shown on the left, together with profile depth. The main diagram shows the occurrence of the primary species, grouped by life form. Latin names for species in *italic script*. Minor occurrences are grouped within the diagram by life form:: 'Other trees' includes *Betula* (birch), *Fraxinus* (beech), *Pinus* (pine), *Sorbus* (rowan), *Tilia* (lime), *Ulmus* (elm); 'Other shrubs' includes *Hedera* (ivy), *Ilex* (holly), *Juniperus* (juniper); 'Other herbs' includes Apiaceae, *Astera* (Aster), *Bellis* (daisy), Brassicarceae, Chenopodiacaea (goosefoot family), *Filipendula* (meadowsweet), *Hypericum* (St. John's wort), *Plantago* (plantains), *Rhinanthus* (rattles), *Rununculus* (buttercup), *Rumex* (docks and sorrels), *Sanguisorbia* (burnet); Spores, e.g. *Polypodium*. To the right is a summary diagram showing the progressive change of the major life forms. Note that *Calluna* (heather), a shrub, is treated here as a dwarf shrub.

2008). Overall, the summary data shown in Figure 2.11 indicate a marked decrease in regional tree and large shrub cover from about 2000 years ago, and a simultaneous increase in *Caluna vulgaris* (common heather) cover, especially in the last 1000 years. The sediments of peat bogs also contain information on changes in surface wetness through time, which for a site such as Archer Moss would reflect climatic conditions. Changes in the decomposition of peat are governed by bog water-table depths, and at Archer Moss the peat properties indicate a number of late Holocene 'wet shifts' in the local climate around 2300, 1500 and 500 years ago. (for a fuller explanation see Chiverrell, 2001; Harvey and Chiverrell, 2004; Chiverrell *et al.*, 2008).

For the later Holocene, the Archer Moss sequence is now augmented by a series of shorter-term, dated pollen profiles from soils and organic deposits buried by alluvial fan and debris cone sediment at the outlets of tributary gullies. These are summarized by Chiverrell *et al.* (2008). The sites include Grains Gill in lower Carlingill (Cundill, 1976, 2000; Chiverrell *et al.*, 2008), Blakethwaite at the head of the Carlingill valley (Harvey, 1996; Chiverrell *et al.*, 2007, 2008), Burnt Gill in Middle Langdale (Harvey *et al.*, 1981; Chiverrell *et al.*, 2008), and Calf Beck in the Chapel Beck valley (Chiverrell *et al.*, 2008). The Grains Gill site (2290 +/-80 BP to 2145 +/-40 BP) and the Calf Beck site (to 2055 +/-35 BP), relate to periods prior to the main woodland decline. Hazel pollen is abundant at both sites, but abundant heather suggests partially open landscapes. Declines in oak and birch through the profiles, with increasing open ground indicators, grasses (*Poaceae*) and sedges (*Cyperacea*), reflect declining woodland cover and the opening up of the landscape around 2200 years ago. The Burnt Gill site (2580 +/-55 BP to 940 +/-95 BP) shows a marked tenth-century decline in tree pollen, whereas the Blakethwaite buried peaty soil itself (1900 +/-60 BP to 1210 +/-50 BP) indicates a partial tree cover at the time (local riparian willows) but abundant heather pollen.

Altogether, the pollen data (Fig. 2.11) suggest a wooded or partially wooded landscape prior to about 2000 BP. Then there is evidence for perhaps spatially variable woodland disappearance to about the tenth century AD, after which open landscapes are characteristic. Almost certainly, this change in the tenth century AD relates to the Norse introduction of sheep grazing on the Fells. The pollen sequence provides the environmental context for the previous section of this chapter, within which a major tenth century phase of gully erosion was identified, but with two minor phases, an earlier one around 2000 BP and at least one later, around 500 years ago (see Table 2.1). The tenth-century phase cannot really be related to any 'wet shift' identified within the peat profiles, but the gully erosion episodes 2200 and 500 years ago do correspond with wetter phases recorded in the Archer Moss peat sequences. However, all three gully erosion phases are more likely to have had either 'human' (land-cover change) or very short-term climatic (i.e., extreme rainfall) causes.

Table 2.1: Summary of erosion-related Howgill radiocarbon dates.

Time Ka BP (approx)	Radiocarbon-dated sites	Geomorphic sequence	Wet periods* and Human events
0	22: 65±35 (0-250 BP) Chapel Beck	youngest Chapel Beck fan	
	21: 275±35 (285-430 BP) Chapel Beck	young Chapel Beck fan	*
	20: 280±35 (430-290 BP) Chapel Beck	young Chapel Beck fan	
	19 300±40 (305-430 BP) Bowderdale	soil below young B'dale fan	
	18: 410±50 (334-514 BP) B'th	soil below youngest fan seds	Population recovery
0.5	17: 600±80 (545-649 BP) B'dale	Young valley floor	Depopulation (eg Black Death)
	16: 694±24 (573-668 BP) Carlingill	peat within low terrace	*
	15: 790±40 (680-735 BP) Bowderdale	B'dale pre -main fan seds	Population increase
	14: 890±50 (738-907 BP) Bth	U buried soil top (humic)	
	13: 930±40 (795-910 BP) Bowderdale	U buried soil (top – humin)	
	12: 940±95 (743-931 BP) Burnt Gill	L'dale, Pre - fan seds	
1.0	11: 1042±22 (960-935 BP) L'dale	Top of soil, under debris cone	
	10: 1090±35 (960-1055 BP) Chapel Bk	Pre - fan sediments	
	9: 1210±50 (1061-1224 BP) B'th	Lower buried soil (top-humin)	Norse colonization
		Pre-main fan sediments	
1.5	8: 1770±60 (1573-1812 BP) B'th	Lower buried soil (top-humic)	*Post-Roman woodland regeneration
	7: 1900±60 (1738-1920 BP) B'th	base of buried soil over debris slide	

2.0	6: 2055±35 (1950-2100 BP) Chapel Bk	top of soil under fan gravs	Romano-British
	5: 2145±40 (2057-2297 BP) Grains G	top of peat under fan gravs	
	4: 2290±80 (2151-2357 BP) Grains G	base of peat, under fan gravs	*
2.5	3: 2580±55 (2503-2768 BP) L'dale	base of soil on young terrace	Iron Age
	2: 2975±85 (3001-3317 BP) B'th	charcoal - base of sequence (pre-dates geomorphic activity)	
>3.0	1: 6070±70 (6763-7005 BP) B'dale	base of peat on high terrace (postdates high terrace sediments)	

Radiocarbon dates given first in Radiocarbon Years BP, then as calibrated dates in calendar years BP (pre 1950), in brackets (calibration after Stuiver *et al.*, 2005).

Red indicates dates prior to burial by fan sediments

Site numbers (22 – 1, youngest to oldest) and sources (a-g, see below):
22: Chapel Beck, valley-side fan, top of buried soil (f); **21:** Chapel Beck/ Long Gill, top of buried soil (f); **20:** Chapel Beck/ White Fell, top of buried soil (f); **19:** Bowderdale/ West Fell, peat below youngest fan (g); **18:** Blakethwaite, soil below youngest fan sediments (b); **17:** Bowderdale, peat over valley-floor sediments (a); **16:** Carlingill, peat within low terrace (d, e); **15:** Bowderdale/ Thickcombe, top of buried soil below fan (g); **14:** Blakethwaite, upper buried soil (b); **13:** Bowderdale, buried soil below fan (g); **12:** Langdale/Burnt Gill, below fan sediments (c); **11:** Langdale/Burnt Gill, below fan sediments (e); **10:** Chapel Beck/Swarth Graves, below fan sediments (f); **9:** Blakethwaite, lower buried soil (b); **8:** Blakethwaite, lower buried soil (b); **7:** Blakethwaite, soil below fan (b); **6:** Chapel Beck/Calf Beck, below fan (f); **5:** Grains Gill, below fan (d,f); **4:** Grains Gill, below fan (d,f); **3:** Langdale/nr Burnt Gill, base of soil on low terrace (c); **2:** Blakethwaite, base of sequence (b); **1:** Bowderdale, high terrace (a).

Key to sources: a) Miller, 1991; b) Harvey, 1996; c) Harvey *et al.*, 1981; d) Cundill, 1976; e) Chiverrell *et al.*, 2007; f) Chiverrell *et al.*, 2008; g) Dunsford, 1998. Human events: modified from Winchester, 1987, 2000.

Wet periods (*) from Archer Moss site (Cundill, 1976; Chiverrell *et al.*, 2008).

CHAPTER 3

The modern geomorphic system

Introduction

There are two important aspects to the modern geomorphic system: hillslope sediment supply from active gully erosion, and erosion and deposition within the stream channels, and particularly how the channels respond to sediment input from the gully systems.

There is modern active gully erosion occurring at a number of sites throughout the Howgills (Fig. 3.1), with gullies cut into glacial boulder clay, whose upper layers have usually been modified by periglacial solifluction processes. These gullies are of a much more limited extent than the tenth-century gully systems described in Chapter 2 (compare Fig. 3.1 with Fig. 2.4).

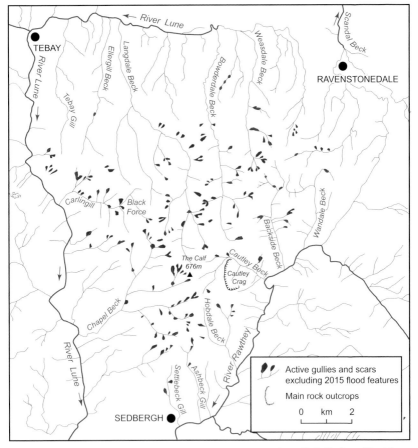

3.1 Distribution of active erosional gullies within the Howgill Fells (modified from Harvey, 1991).

The stream channels are influenced (i) by flood power, controlled by drainage area and valley-floor gradient, itself reflecting the late Pleistocene to Holocene incisional history (see Chapter 2), and (ii) by sediment supply from upstream, but also locally, especially from eroding gullies. Where stream power is high in relation to sediment supply the channels of both the main rivers and the Howgill streams are cut directly into bedrock (Fig. 3.2a). Elsewhere the channels are 'alluvial' channels of two types (Figs 3.2b,c) within Holocene floodplains. The most common type is the single-thread, relatively stable channel (Fig. 3.2b), with well-developed pool and riffle sequences on the gravel stream bed. During moderate to high discharges gravel sediment moves from riffle to riffle, but the channel configuration remains relatively stable. Only during very high discharges does the channel plan view change much, by bank erosion and channel migration (or avulsion). However, the second type of alluvial channel occurs near zones of coarse (i.e. gravel) sediment input, at some tributary junctions, but especially below active gully systems. There the channels flip to the wider, steeper, laterally migrating mode, and in extreme cases to a braided regime (Fig. 3.2c).

The linkage between the gully systems and channels (i.e., hillslope-to-channel coupling) is of fundamental importance and is a dual process. On the one hand, stream erosion undermines the base of the slope, causing a local slope failure. The resulting scar may then extend headwards to form a gully. If the stream then migrates away from slope base, the sediment derived from the gully may be deposited as a fan 'buffering' (i.e., disconnecting) the system. On the other hand, an active gully system feeds coarse sediment to the stream. Channel response to large localized volumes of coarse sediment input (below gully sites, and perhaps some tributary junctions), is channel widening, steepening, and in extreme cases forming a braided channel.

These relationships constitute the 'normal' situation in response to the 'normal' range of annual rainfall triggers and 'normal' stream hydrology (i.e., reaching near-bankfull flood conditions up to several times per year). Times of maximum sediment flux, coincident with near bankfull conditions, similarly would be expected to occur up to several times per year. However, occasional extreme events may overload the system, triggering new gully formation and massive sediment input, to which the channels respond by dramatic channel changes. Two such events have occurred during the 40+ years of monitoring the Howgill systems. In June 1982 a major storm and flood, with a return period estimated c.100 years occurred, which especially affected the northern Howgill valleys (Harvey, 1986; Wells and Harvey, 1987 – see below and Chapter 6). In December 2015 an event of a similar magnitude occurred, from very different meteorological conditions. It is too soon to incorporate the results of any detailed analysis of that event here, but the effects are currently being studied by Professor Richard Chiverrell of the University of Liverpool. A short summary is given at the end of this chapter. It is interesting that two such events (with very different geomorphic responses), of which even the lower (1982) event had a long-term recurrence interval estimated to be in the order of

3.2 Howgill stream channel types: a) bedrock channel; Langdale Beck; b) single thread, relatively stable alluvial channel, Bowderdale Beck; c) unstable braided channel: Langdale Beck, following the 2015 flood.

3.3 Grains Gill gully site, illustrating progressive change: **a)** 1969 – erosion active;
b) 1985 – stabilization initiated; **c)** 2007 – progressive stabilization.

100 years, and the larger (2015) event with an even longer recurrence interval, have occurred within less than 40 years under very different meteorological conditions. Might this be an indication of current climatic change?

Gully processes

I have monitored the geomorphic system in the Carlingill valley, especially at the Grains Gill gully site (Figs 3.3, 3.4) over a 40-year period (Harvey,

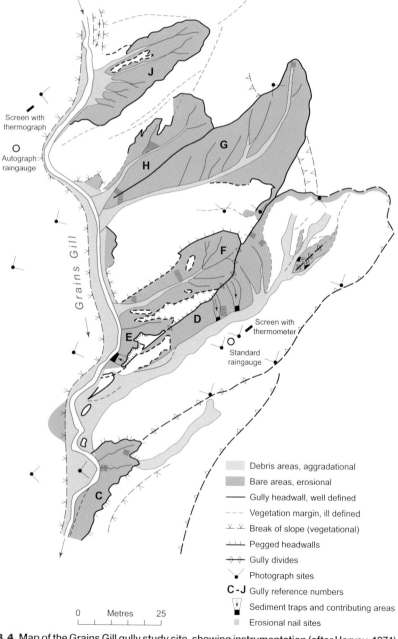

3.4 Map of the Grains Gill gully study site, showing instrumentation (after Harvey, 1974).

1974; 1977, 1987ab, 1991, 1992, 1994; Harvey *et al.*, 1979; see also Chapter 5), at a range of spatial and temporal scales. At the most detailed scale I installed sediment traps (large box traps) at the base of two gullies, and every 10 days to two weeks over a period of 2 years (1971–1972) I collected and weighed the accumulated sediment. I had set up a field laboratory, involving an autographic raingauge, a thermograph and simple sediment analysis equipment (not much fun working in (frequent) wet or cold weather!). I had also set up open (chicken-wire) traps on unrilled slopes to trap coarse sediment only, which I also operated subsequently for a further two years (1975–1977). At the same time as the traps were set up, I set up a longer-term (30+ years) less intensive monitoring that involved measuring pegged headwalls to record headwall recession rates, and erosion pins on the gullied slopes to measure surface erosional lowering. These measurements, together with sequential photography, I maintained at six-monthly intervals (spring and autumn) to record seasonal variations as well as longer-term morphological change. The photographic cover was also maintained, though less intensively at other gullied sites in the Carlingill valley (see later; Chapter 5, see also Harvey and Calvo-Cases, 1991).

The results of the shorter-term intensive monitoring demonstrate a clear seasonality in sediment yields. This monitoring revealed a seasonality of dominant process, which is in turn expressed in a seasonality of morphology (Harvey, 1974; 1987a). Despite similar seasonal precipitation totals, winter sediment yields are much heavier and much richer in stones (>25mm diameter) than summer yields (Table 3.1; the balance is also shown graphically in Fig. 3.5). The seasonality reflects a dominance of mass-movement processes (mudflows) in winter, influenced by freeze-thaw conditions and snowmelt as well as by rainfall, as opposed to runoff processes alone during the summer. Summer runoff causes winnowing of fine material, derived especially from seasonally incising rills. The seasonality is in turn reflected in the morphology of the gullied slopes, with the rill network well developed during the summer months, but largely destroyed during the following winter (Fig. 3.6). A similar seasonality has been observed in other environments, for example in New Jersey (USA) (Schumm, 1956). We have also observed a similar phenomenon in Mediterranean Spain, but with the seasonality reversed (Harvey and Calvo-Cases, 1991).

Over the longer term, several trends are apparent. Gully headwalls are receding. From the headwall measurements made over the period 1969–89 the recession rate (E, m/yr) can be related to gully length (L, m) by the regression equation:

$$E = 0.94 \, L^{-0.51} \ (n = 10, \ r = -0.61) \ \text{(Harvey, 1991)}.$$

With only 10 data points this relationship is statistically significant only at the 10% level, but does give a crude indication of how recession rates decline with gully age. A second trend is the recolonization of basally stable gullied slopes by vegetation, which tends to colonize from the slope base upwards. It is difficult to quantify this process, but at three sites the upslope advance

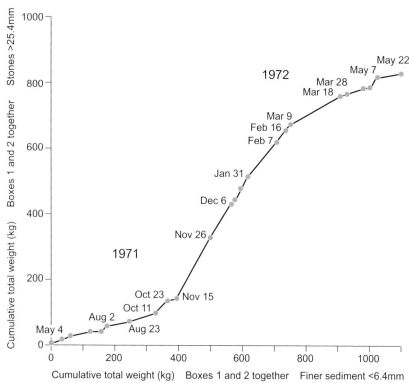

3.5 Seasonal particle-size characteristics of sediment yield by gully erosion at the Grains Gill study site (1971–1972), illustrated by a double-mass curve between total yield and yield of stones >25.4mm diameter (after Harvey, 1974).

of a vegetation front was measured over the period 1969–87 with remarkably consistent rates of 0.35–0.4 m/yr (Harvey, 1991). Putting these two rates together it is possible to develop a simple model of gully growth and stabilization (Fig. 3.7; see also Harvey, 1992). On this basis we might estimate that the 'lifespan' of individual gullies in Carlingill could be up to 150 years. This is almost certainly an over-estimate, which is difficult to test, given the uncertainty of the age of the modern gullies. The gullies are recognizable on the (small-scale) air photos taken in 1948, on which bare, unvegetated areas can be identified. However, within the last 20 years the rate of revegetation of stabilizing slopes has clearly accelerated (see Fig. 3.3). If the revegetation rate was exponential rather than linear, the lifespan of individual gullies would be about half that estimated by the linear model (see Fig. 3.7). The model, in any case, assumes stable basal conditions. It would not apply to gullies based by an active stream. Such gullies tend to produce debris cones at the gully base, which could be periodically removed by stream floods, thus rejuvenating the gully upslope. This introduces a third trend, specific to gullies based by active streams: the cyclic build-up and removal of debris cones (Fig. 3.8a,b). In other words, slope-stream coupling is important in influencing both sets of processes (see below).

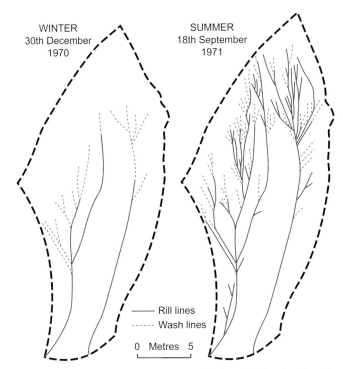

3.6 Seasonality of rill networks: comparison of winter 1970 network with summer 1971 network on Gully D at Grains Gill gully site (for location see Fig. 3.4) (modified from Harvey, 1992).

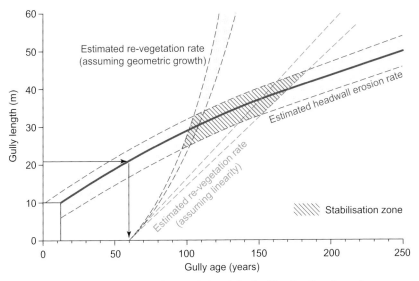

3.7 Gully growth model for the Grains Gill and Carlingill gully sites, based on measured headwall erosion rates and measured vegetation encroachment rates. This model assumes a linear rate of vegetation encroachment (see green zone on the diagram). Stabilization would be accelerated if the rate of vegetation encroachment itself accelerated (see red zone on the diagram). The model does not take into account renewed activity related to basal scour (modified from Harvey, 1992).

3.8 The effects of basal scour and debris cone removal on gully erosional status: Grains Gill, Gully C (see location on Fig. 3.4). **a)** Pre-scour situation; **b)** after basal scour.

Stream channels
(processes and morphology)

Before considering slope-stream coupling, it is necessary to consider the range of alluvial channel morphologies characteristic of the Howgills. The two types of alluvial channel have already been defined: single-thread, relatively stable channels (Fig. 3.2b) and wider, shallower, in extreme cases unstable braided channels (Fig. 3.2c). In Carlingill, prior to the December 2015 floods, there had been only minor changes in the configuration of the single thread channels since the 1948 air photos, but the braided reaches have been characterized by instability. Using lichenometry, we have reconstructed the sequence of channel changes in the upper part of the Carlingill valley over the last *c.*150 years (Harvey *et al.*, 1984; see Chapter 2 above, see also Fig. 3.9). Two tributary streams feed into this reach, together with long-standing gullies and scars cut into solifluction-capped glacial till as well as into gravels of the high terrace. The reach is quite clearly one of chronic instability.

As would be expected channel morphology, expressed by channel width (w, m) and channel slope (s), of the single thread channels for Howgill streams (the data derived primarily from Langdale and Bowderdale), reflects the normal flood discharge regime, and therefore can be related to drainage area (A, km^2) by the following regression equations (Harvey, 1987b)(see also Fig. 3.10):

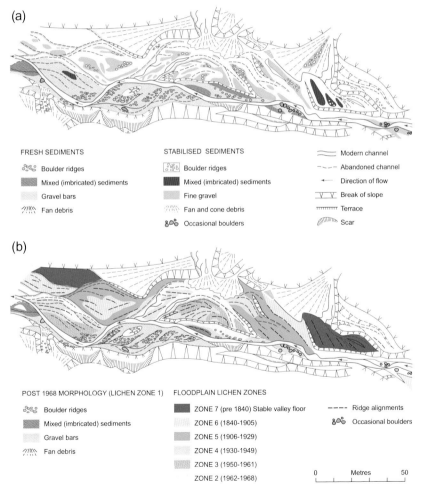

(a)

FRESH SEDIMENTS

- Boulder ridges
- Mixed (imbricated) sediments
- Gravel bars
- Fan debris

STABILISED SEDIMENTS

- Boulder ridges
- Mixed (imbricated) sediments
- Fine gravel
- Fan and cone debris
- Occasional boulders

- Modern channel
- Abandoned channel
- Direction of flow
- Break of slope
- Terrace
- Scar

(b)

POST 1968 MORPHOLOGY (LICHEN ZONE 1)

- Boulder ridges
- Mixed (imbricated) sediments
- Gravel bars
- Fan debris

FLOODPLAIN LICHEN ZONES

- ZONE 7 (pre 1840) Stable valley floor
- ZONE 6 (1840-1905)
- ZONE 5 (1906-1929)
- ZONE 4 (1930-1949)
- ZONE 3 (1950-1961)
- ZONE 2 (1962-1968)

- Ridge alignments
- Occasional boulders

0   Metres   50

3.9 Surveyed maps of the middle reach of Carlingill (see also Fig. 2.9), showing: **a)** sediment properties on the valley floor; **b)** lichenometric dating of the features shown above, based on the calibration curve for the lichen *Rhyzocarpon geographicum* (see Fig. 2.3) (modified from Harvey *et al.*, 1984).

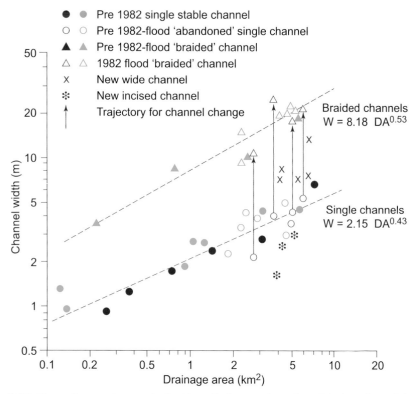

3.10 Channel geometry graphs for Howgill streams, based on data from Langdale (shown in blue) and Bowderdale (shown in brown), showing width to drainage area relationships for single thread (relatively stable) alluvial channels, and for unstable braided channels. Included are the effects of channel change that resulted from the 1982 flood (see later; see also Harvey, 1987b).

$$w = 2.15\ A^{0.43}\ (n = 24, r = 0.93)$$
$$s = 0.040\ A^{-0.52}\ (n = 24, R = -0.92)$$

If coarse sediment calibre (cm) is also taken into account using multiple regression analysis, these correlation coefficients improve to 0.97 and 0.98 respectively. In Figure 3.10 the plotting positions for the braided reaches, for the width relationship, stand well above those for the single thread reaches. The regression equations for the braided channels, including those that braided as a result of the 1982 flood, are as follows:

$$w = 8.18\ A^{0.53}\ (n = 14; r = 0.92)$$
$$s = 0.073\ A^{-0.75}\ (n = 14, R = -0.97)$$

Also shown on the graph on Figure 3.10 are the effects of the 1982 flood on the channels of Langdale and Bowderdale Becks, where some reaches, prior to the flood were single thread, but during the flood switched to braiding, and as a result accorded with the width relationship for braided channels.

Coupling relationships

We now need to consider the nature of the coupling relationships, as these control both gully development and channel regime. They are fundamental

3.11 The significance of gully to channel coupling relationships illustrated from Greencomb Gills, Bowderdale. The gully on the left fed sediment to an alluvial fan at its base and therefore became discoupled from the basal stream (Bowderdale Beck). Since then the stream channel has eroded the base of the fan, allowing the tributary gully to trench through the fan and supply its sediment directly into Bowderdale Beck – that system is now coupled. In contrast, the gully on the right is NOT coupled to Bowderdale beck – all its sediment accumulates as an alluvial fan on the valley floor.

for the dynamics of the Howgill geomorphic system, not only in the context of the modern gully–channel relationships but also over the longer term (Harvey, 1994, 1997, 2001). The significance of coupling is illustrated on Figure 3.11, where on the left of the photo is a small tributary to Bowderdale Beck (Greencomb Gill, see Chapter 6B) which is footed by a small, previously stable alluvial fan (probably related to the tenth- century gullying phase). Bowderdale Beck has now undercut the fan, causing the tributary stream to incise and deliver its sediment directly into Bowderdale Beck. In contrast, a small tributary on the right of the photo which has no connection (is therefore non-coupled) with Bowderdale Beck is depositing its sediment on a small actively aggrading alluvial fan.

Coupling relationships have significance far beyond the Howgills! – at a wide range of scales within geomorphology. They control the sediment supply to and through the fluvial system (Harvey, 2000, 2002) and are especially important in the geomorphology and dynamics of alluvial fans (Harvey, 2012). A summary of the timescales involved in the Howgill coupling relationships is given in Table 3.2, ranging from the 'normal' slope/stream coupling, described above, through the effects of extreme events, described below, to the long-term effects of 'total overload' as in the tenth-century situation described in the previous chapter. It is interesting to compare the coupling relationships of the modern gully-to-channel system with those of the tenth-century system. The modern gullies are basally induced and coupling is a two-way process.

Basal scour by the stream accelerates gully erosion. Gully-derived sediment is fed into the channel, but temporarily buffers the system by the formation of basal cones, only to be cleared by floods, thereby re-activating erosion within the gullies. For the tenth-century system coupling is primarily a one-way process, in that the large gully systems were not affected by processes within the main stream channels. The large gully systems were activated by conditions on the slope rather than by those in the basal stream. These gullies fed sediment either directly into the stream or into large alluvial fans at the junction between gully channel and the main stream. The fans had primarily a one-way coupling relationship with the main stream, but only over a very long term could the effects of toe erosion of a large fan be transmitted upslope to the gully systems.

Effects of extreme events (1982 and 2015)
A) The flood event of June 1982
On 6 June 1982 an exceptionally heavy convectional storm hit the northern Howgills (Harvey, 1986). The storm was one of a series of thunder cells that affected northern England on that day. Rainfalls recorded at local raingauges were very variable (for example 50mm at Grayrigg, west of the Howgills, but only 11mm at Orton, north of the Howgills), and are of little use in estimating the rainfall over the Howgills themselves. Of much more use is The Meteorological Office radar data, which indicate 55–65mm for the two 5km x 5km grid squares that cover the central and northern Howgills. Local witnesses give the timing of the storm from 13.00 GMT to 15.30 GMT, but with exceptionally heavy rain during the first 45 minutes of the storm. Using the models from the National Flood Studies Report (NERC, 1975), the return period for such a storm in this area would be of the order of 100 years. The intensity of the storm is confirmed by the timing of the flood runoff recorded at the Tebay measuring station on the River Lune. At 1500 GMT river flow rose from $c.3m^3$ sec^{-1} to a peak of $124m^3$ sec^{-1} in less than three hours, rapidly falling back to normal by the following morning. This was clearly a 'flash-flood' situation.

The geomorphic effects of that storm were dramatic, especially in Langdale and Bowderdale (Wells and Harvey, 1987; see also Chapter 6). Slope failures and fresh erosional scars were widespread, especially within the tenth-century gully systems (Fig. 3.12, see also Harvey, 1986). These fed sediment towards the main valleys, 'overloading' the coupling capacity, resulting in the deposition of enormous fans at the gully foots or in tributary junction situations (Harvey, 1987b), burying the older tenth-century fans and cones. Within Langdale and Bowderdale there were thirteen such cones (Table 3.3), which we grouped into four classes, based on the sedimentology of their constituent deposits. The sedimentology ranged from cohesive debris flows, composed of ill-sorted matrix-supported material, through transitional deposits of looser, ill-sorted stony sediments, to true imbricated fluvial gravels, including both coarse bar forms and finer sheet forms. We based our primary interpretation of the facies on depositional morphology, clast size and fabric. This interpretation was backed up by our laboratory analysis of the particle-size characteristics of

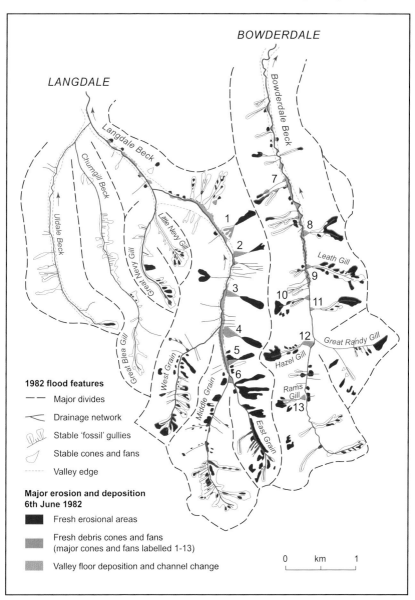

3.12 Erosional and depositional effects of the 1982 flood on the morphology of Langdale and Bowderdale (modified from Harvey, 1986). Numbered fans and cones as follows – Langdale: 1. Thrush Gill; 2. East Combs Gill; 3. Wolflea Gill; 4. West Hazel Gill; 5. New Gill; 6. East Grain – Bowderdale: 7. Greencomb Gills; 8. Lodge Gill; 9. Leath Gill; 10. Woofler Gill; 11. Wotey Gill; 12. Hazel Gill; 13. Rams Gill. Dominant facies: Cohesive debris flows (D) – Fans 1,5,7; Transitional deposits (T) – Fans 2, 3, 8, 10; Coarse fluvial bar sediments (F1) – Fans 4, 9, 11; Fine fluvial sheet sediments (F2) – Fans 6, 12, 13.

the sediment matrix (Fig. 3.13; Table 3.4). We interpreted these four facies to reflect the fluidity of the water:sediment mix supplied (Wells and Harvey, 1987). We identified four cones as 'type' examples, each dominated by one

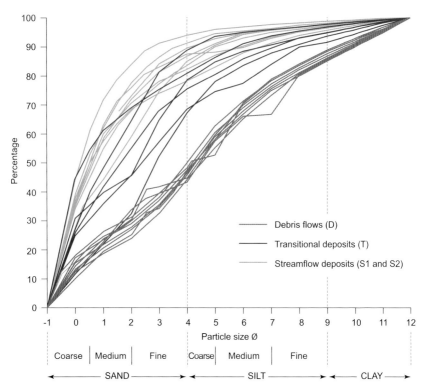

3.13 Matrix grain-size distributions for 24 sediment samples of the 1982 deposits from Thrush Gill (Langdale), Lodge, Leath and Hazel Gills (Bowerdale), for the major facies types: Debris flows (Facies D), Transitional deposits (Facies T) and Streamflow deposits (Facies S1, S2) (modified from Wells and Harvey, 1987). Particle size is shown in ph1 (Φ) units, a negative logarithmic scale, the conventional scale used for sediment sizes (see Appendix 2).

of the four sedimentary facies (Fig. 3.14). The morphology of the cones and fans reflects the dominant mode of deposition. Debris flows from the smaller steeper catchments were deposited on steep cones but low-angle fluvial fans were formed from the larger catchments (Fig. 3.15). On any one fan or cone, there was dominance of one type of sediment, but each fan or cone also showed a within-storm depositional sequence, apparently reflecting within-storm variations in the water/sediment mix (Fig. 3.16).

The channels of Langdale and Bowerdale Becks underwent at first severe erosion, then massive deposition. In a number of locations the sediment overload caused switching from single-thread to braided mode (Fig. 3.17).

The storm effects clearly disrupted the 'normal' coupling characteristics described above. Over the years that followed, the Langdale and Bowerdale systems gradually returned towards the 'normal' situation (Harvey, 1991; 2007). The erosional scars began to heal by revegetation, and the fans and cones also began to stabilize by revegetation. The fluvially dominant fans underwent some dissection; and some of the 1982 braided channels reverted to single-thread status (Figs 3.18, 3.19).

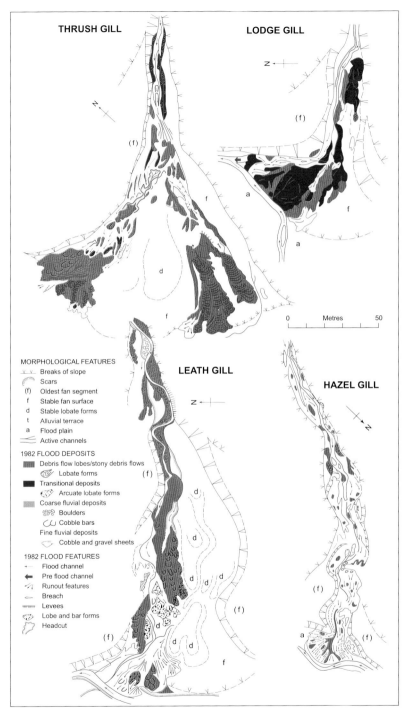

3.14 Maps of debris-cone/alluvial-fan deposition from the 1982 flood in Langdale and Bowderdale, illustrating the four facies-types of deposition. Thrush Gill (Langdale) is dominated by cohesive debris flows, Lodge Gill (Bowderdale) by stony debris flows, Leath Gill (Bowderdale) by coarse fluvial sediments and Hazel Gill (Bowderdale) by fine fluvial gravels. (after Wells and Harvey, 1987).

3.15 Morphology of selected cone and fan deposition during the 1982 flood: **a)** Thrush Gill (Langdale), dominated by cohesive debris flows – note lobate topography and levees; **b)** Woofler Gill (Langdale) dominated by low-angle fluvial-gravel sheets.

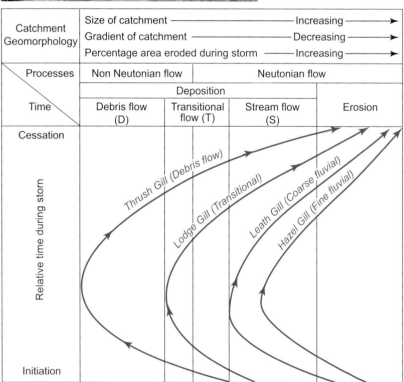

3.16 Interpretation of the within-storm sedimentary sequences within the four type-examples of debris cones (modified from Wells and Harvey, 1987).

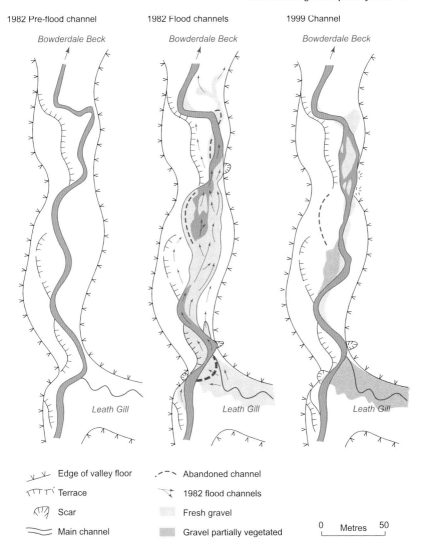

1982 Pre-flood channel   1982 Flood channels   1999 Channel

Bowderdale Beck   *Bowderdale Beck*   *Bowderdale Beck*

Leath Gill   *Leath Gill*   *Leath Gill*

Edge of valley floor

Terrace

Scar

Main channel

Abandoned channel

1982 flood channels

Fresh gravel

Gravel partially vegetated

0   Metres   50

3.17 Channel change in response to the 1982 flood: the example of Middle Bowderdale. **a)** The 1982 pre-flood channel; **b)** the 1982 flood channel; **c)** recovery by 1999 (modified from Harvey, 2007).

3.18 Cone recovery from 1982 flood: East Grain cone a) in 1982; b) in 2007.

(a)

(b)

3.19 Channel recovery from 1982 flood; Middle Langdale **a)** in 1982; **b)** in 2007.

Another major storm event occurred in December 2015, affecting Carlingill as well as the Langdale and Bowderdale valleys. It is too early to present a detailed analysis of the effects of that storm and flood, but Professor Richard Chiverrell of the University of Liverpool is currently working on the data. We present a preliminary report below.

B) The flood event of December 2015
(Preliminary Report written in co-operation with Richard Chiverrell **)

The December 2015 flood was a very different flood event from that of July 1982. First, it was a winter rather than a summer event, and was the culmination of a severe wet period rather than the effect of a single intense convectional storm. November 2015 had been the second wettest November in Cumbria since 1910. The third of December was a very wet day, with over 40mm of rain recorded at Shap, to the NW of the Howgills. Then on 5 December a major rainfall event occurred over Cumbria, with Shap recording 108mm. The effects were devastating. Flooding and landslides blocked road and rail routes throughout Cumbria. The rivers recorded the highest discharges for many years. The Lune at Tebay rose from an already high mean daily flow of $77m^3 sec^{-1}$ on 4 December to a peak of $409m^3 sec^{-1}$ on 5 December. This was a lot higher than the flood of 6 June 1982, and therefore would have an even longer return period. Fairly high flows (above $c.20m^3 sec^{-1}$) persisted for several days, with a secondary peak of $162m^3 sec^{-1}$ on 9 December. Later in December there were three further flood peaks in excess of $150m^3 sec^{-1}$ as a series of cyclonic pressure systems crossed the British Isles. Within the Howgills the main streams all had very high flows, not only on 5 December, but on several later occasions during the month. Presumably from the very high peak on 5 December, flood debris was left several metres above normal flow levels, especially within confined reaches.

The geomorphic effects were considerable, but differed markedly from those of the 1982 flood. Quite clearly the effects were dominated by channelized floodflow rather than by hillslope erosion. There were a few small slope failures on open slopes, but most slope failures were associated with the incision of small hillslope gully streams. In comparison with the effects of the 1982 event there was remarkably little erosion within the large (*c*. tenth century) hillslope gully systems. Almost all fresh erosional scars or gully reactivations were basally induced in response to main stream floods, particularly in Langdale, Bowderdale and Carlingill, apparently less so in the other Howgill valleys. The Grains Gill tributary to Carlingill (see above) was remarkably little affected. Two scars in lower Grains Gill underwent significant basal erosion, but in the now largely stabilized Grains Gill gully system there was only minor erosional activity. Along Carlingill itself two streamside scars/gully systems underwent major basal erosion, at one site (Fig. 3.20a) triggering major mudflows that blocked and diverted Carlingill across its floodplain. Since then (by late 2016) the channel has incised itself into the floodplain to the south of its former position.

Further upstream there were other, mostly streamside, sites that underwent erosion. At one site (in the vicinity of Gully U – for location see Chapter 5), there was a shallow slope failure in mid-slope on the flanks of Uldale Head on the northern margin of the Carlingill valley. The resulting debris flow travelled down to the valley floor and was incised by the later December

3.20 Examples of the effects of the 2015 flood. **a)** Gully M in Carlingill (compare with Fig. 5.6b). During the flood the stream undermined the gully system, causing huge debris flows to inundate the base of the gullies, forcing the stream to migrate towards the south across the floodplain (to the right of the photo). **b)** Widened channel in Langdale, showing marked bank erosion cutting into floodplain and low terrace sediments, and extensive gravel deposition on the inside of the resultant bend.

floods, but has since been decoupled from the stream and has begun to stabilize by revegetation (for more detail see Chapter 5).

On the valley floor the stream channel in many places underwent major bank erosion into either floodplain or low terrace sediments, resulting in significant channel widening, and in some cases in braiding. This transformed the channel width relationships for much of Carlingill from those dominantly of stable single thread channels to those of unstable laterally migrating or braided channels (To correspond with the upper relationship shown in Fig. 3.10).

The effects described for Carlingill apply also in Bowderdale and especially in Langdale. In complete contrast with the 1982 storm, there were only very minor reactivations in the headwater gully systems within these two catchments. As a result there was almost no new sediment deposited on the fans and cones at sites of the tributary junctions with the main streams. There was erosion of a number of streamside scars, particularly in Langdale. However, the main channels underwent major bank erosion of both floodplain and low terrace sediments, resulting in dramatic channel widening and at many sites (Fig. 3.20b), especially in Langdale, with the channel switching to a new braided regime.

The overall effects of the 2015 flood contrast markedly with those of the 1982 storm. For the 1982 storm the intense rain triggered major slope erosion, especially within the previously mostly stabilized gullied catchments. The resulting massive sediment input dominated the downsystem response, causing major sedimentation on the tributary-junction fans and cones, and major sediment input into the main channels. The channels responded at critical sites by switching from a stable single-thread mode to an unstable migrating, and in extreme cases, a braided mode. Those effects persisted and were still evident in the landscape 35 years later. In contrast, the effects of the December 2015 floods were dominated by the very high stream flood discharges. Erosion of the slopes was dominantly basally induced, with the one exception in the Middle Carlingill valley, described above. Where the channels impinged on the valley sides there were basally induced slope failures and relatively small-scale gullying effects. Major erosion took place on the valley floors, of floodplain or low terrace sediments, causing rapid channel widening and lateral migration. The high sediment input from these sources transformed many reaches of Langdale and Bowderdale Becks especially, which switched from stable, single-thread mode to unstable, lateral migratory and even braided mode, sustained for considerable distances downstream. Essentially what has happened is the morphological transformation of the valley floors, producing a morphology that is likely to persist for a considerable time.

**Professor Richard Chiverrell: Dept of Geography and Planning, School of Environmental Sciences, University of Liverpool, PO Box 147, Liverpool L69 3BX.

Table 3.1: Grains Gill gullies – seasonal sediment yield (kg) collected in box traps on rilled slopes, and open traps on unrilled slopes (Harvey, 1987a,b).

Season	Rainfall (mm)	Frost (0 days)	Sediment yield (kg)		Open Traps
			Box Traps		
			Total	Stones	Stones
Box traps					
Summer 1971	604	0	514	102	-
Winter 1971–72	758	138	1817	688	238
Summer 1972*	349	0	507	106	6
Open traps					
Winter 1975–76	514	97			154
Summer 1976	600	0			50
Winter 1976–77	571	149			362

*3-month period only

Table 3.2: Coupling relationships within the Howgill geomorphic system: timescales.

Process	Event frequency	Response/Recovery time
On-slope gully erosion	30–35 events per year	Seasonal cone development
Gully basal scour	2–6 years	Channel response: c.5 years
Basal gully initiation	1 event within c.40 years	Gully stabilization: c.150 years
Threshold event	c.100 years	Channel recovery: c.20–40 yrs
Total overload (e.g. 10th c. AD)	Once in 5ka	Overall stabilization c.300–500 years

Table 3.3: Sedimentology and morphology of the debris cones/alluvial fans deposited by the June 1982 flood event in Langdale and Bowderdale.

Catchment	Area (km²)	Catchment slope	Dominant facies	Depositional gradient (fan slope)
Langdale				
1.Thrush Gill	0.056	0.56	D	0.39–0.25
2. East Combs Gill	0.063	0.53	T	0.38–0.28
3. Wolflea Gill	0.088	0.34	T	0.25–0.18
4. West Hazel Gill	0.061	0.42	S1	0.37–0.18
5.New Gill	0.024	0.50	D	0.53–0.39
6. East Grain	0.794	0.21	S2	0.12–0.10
Bowderdale				
7. Greencomb Gills	0.040	0.47	D	0.33–0.09
8. Lodge Gill	0.135	0.39	T	0.26–0.18
9. Leath Gill	0.331	0.36	S1	0.19–0.09
10. Woofler Gill	0.126	0.35	S2	0.12–0.10
11. Wotey Gill	0.119	0.38	S1	0.12–0.10
12. Hazel Gill	0.218	0.31	S2	0.14–0.11
13. Rams Gill	0.113	0.40	S2	0.15–0.12

For locations – see numbered catchments on Figure 3.12.
Catchment slope: Catchment relief/catchment length.
Dominant facies: D – Cohesive debris flows; T – Transitional stony debris flows;
S1 – Coarse fluvial bars; S2 – Finer fluvial sheet gravels.

Table 3.4: Characteristic particle-size percentage ranges for the finer (matrix) fractions (i.e., sand, silt, clay) for three facies groups (debris flow deposits – facies D; Transitional deposits – facies T; and streamflow deposits – facies S1,S2), for 24 sediment samples taken from the 1982 debris cones/fans in Langdale and Bowderdale.

	Characteristic percentage ranges for:		
	sand	silt	clay
Debris flow deposits (Facies D)	45–60	35–45	10–20
Transitional deposits (Facies T)	70–90	10–30	5–15
Streamflow deposits (Facies S1, S2)	80–95	5–20	0–5

Part Two
The Field Sites

Introduction

The following chapters present the field sites. *Locations, instructions, etc. will be given in italic script.* A word of warning: there are no roads within the Howgills, so with the exception of the reconnaissance tour described in this chapter, for all the other field excursions (described in Chapters 5–7), you need to park on the margins of the fells. Then, all field visits are on foot. Note that in most cases the country is rough, sometimes involving steep or treacherous slopes. Some streams will need to be crossed by wading. In normal situations this should not be a problem. Take care, and be prepared to modify your route if streamflows are high. Hiking boots are essential; raingear is also a sensible precaution! Approximate distances for each field trip are given in kilometres; locations are given in National Grid references and in GPS coordinates (see also Appendix 3).

Langdale Beck, central Howgills: braided river channel (see Chapters 3, 6).

CHAPTER 4

A reconnaissance road trip around the Howgills

The purpose of this excursion is an introduction to the geomorphology of the Howgill Fells, seen from outside the Howgills themselves. It is the only excursion presented here that is essentially by car. All the others are walking trips into the interior of the Howgills.

*As this is essentially a circular tour, (for the route map see Fig. 1.1) the route could be picked up at any point within the circuit. For convenience this guide is written assuming a start from Sedbergh; alternatively details of an approach from Kendal along the A685 (Kendal to Tebay and Kirkby Stephen road) are also given. From Sedbergh take the A684 west of the town. After about 2.5km, a little after crossing the River Lune, turn right onto the B6257 towards Tebay. Continue along that road. Note the views to the east, across the River Lune towards Chapel Beck and the south-western Howgills (similar to that shown in Fig. 1.9). Continue until the road meets the A685 at Dillicar Common, turn right onto the A685 and continue north to just over the brow to a track on the left where it is possible to pull in (**Stop 4.1**). Alternatively, leave Kendal by the A685 towards Tebay and Kirkby Stephen. Drive for about 11km to Dillicar Common (about 3.5km beyond Grayrigg)* [SD 613988: 54.3861, -2.5955] *where the B6257 from Sedbergh joins the A685, then go as directed above to **Stop 4.1**. From here, on the right of the road there is a view of the Howgill Fells to the east, including the Carlingill catchment.*

Stop 4.1 Dillicar Common [SD 613988: 54.3861, -2.5955]
From here you are looking north into the Lune Gorge (see Fig. 1.3), the gorge cut by the southward glacial (?) diversion of the Lune drainage (see Chapter 1). Also from here there is a view to the east across the Lune Valley into the Carlingill catchment (Fig. 4.1a) within the western Howgills (see Chapter 5). The plateau-like summit surfaces of the tops of the Howgills are probably remnants of the Late (?) Tertiary erosion surfaces (see Chapter 1). The Carlingill valley is cut deeply below these surfaces. The valley sides are mantled by Late Pleistocene solifluction deposits, below which the modern Carlingill valley is deeply entrenched. To the left (north) of the Carlingill valley is the Grains Gill tributary, within which the Grains Gill gully system is visible (see Chapters 3, 5). Full details are described in Chapter 5. Below you the Lune valley is deeply entrenched. The modern valley (Fig. 4.1b) includes Late Pleistocene to Holocene river terraces (we do not know the details) and the modern gravel-bed alluvial river channel set within a Late Holocene floodplain.

*Continue north along the A685 for about 1km into the Lune valley downstream of the Lune Gorge, where there is a parking area and a scenic overlook on the right, over the lower part of the Lune Gorge. Park here. This is **Stop 4.2**.*

4.1 The western Howgills: a) general view from Dillicar, including the Carlingill valley (see Chapter 5); b) terraces of the River Lune downstream of the Carlingill confluence.

Stop 4.2 Lune Gorge overlook [NY 607007: 54.4005, -2.6053]

This site gives an excellent overview of the southern end of the Lune Gorge and the western flank of the Howgills. To your south across the motorway the low Holocene terraces of the Lune are evident. Beyond and to the southeast is the Carlingill valley, which you saw from the previous stop. Swinging your view round to the left, immediately east of you across the valley is Blease Fell and the small, steep, gullied catchment of Blease Gill (Fig. 4.2a). The gullies, now largely stabilized, are probably of tenth century (Viking) origin, as are most similar valley-head features in the Howgills. Move your gaze around to the north. You are now looking into the Lune Gorge itself (see Fig 1.3;

4.2 The Lune Gorge area (see also Fig. 1.3). **a)** View across the southern part of the Lune Gorge to the western flanks of the Howgills at Blease Fell; note the (now mostly stabilized) gully system at the head of Blease Gill. **b)** Cirque-like form on the SW flank of the Lune Gorge; **c)** Gravel-bed channel of the River Lune at Borrowbridge, downstream of the Lune Gorge; **d)** Bedrock channel of the River Lune at Tebay within the Lune Gorge.

discussed in relation to the next stop). To complete your visual circuit, look now behind you to the W and SW. You are looking into one of the two crude cirque-like erosional forms that flank the west side of the Lune Gorge (Fig. 4.2b). These two forms may have supported small short-lived cirque glaciers during the closing stages of the regional glaciation, or at least they are nivation cirques, produced under periglacial conditions during the final, closing stages of the last glaciation.

*Return to the A685 road; turn right (north) across the Borrow Beck tributary of the Lune. Then you come to a sweeping curve across the M6 motorway, the west coast main railway line and the river Lune, before you come to a junction on the left with the old road, now a dead end. Turn left here into the old road and park. This is **Stop 4.3**, Lune's Bridge* [NY 613028: 54.4197, -2.5979].

Stop 4.3 Lune's Bridge [NY 613028: 54.4197, -2.5979]

You are now in the central part of the Lune gorge. I suspect that the Lune Gorge is the result of the glacial diversion of the River Lune southwards from its original course north of the Howgills, rather than the simple product of regional base-level change (for discussion of this theme see Chapter 1). Behind you to the southwest are the crude cirque-like forms that may have been the products of later glacial erosion, or at least are nivation cirques, produced under periglacial conditions during the closing stages of the last glaciation. Now-stabilized and mostly vegetated periglacial screes also mantle the steep slopes on the east side of the valley.

The River Lune downstream from here has a cobble-bed alluvial channel (Fig. 4.2c), but a little upstream within the Lune gorge itself it is cut into bedrock (Fig. 4.2d), the Upper Silurian, Bannisdale slates (see Chapter 1). This bedrock reach represents an important knick point in the profile of the modern River Lune, but rather than being induced by regional base-level change I suspect it is a direct result of the glacial diversion of the Lune south into its present course through the Lune Gorge (see Chapter 1).

Return to your car. Return to the A685 main road and turn left into Tebay village. In Tebay village [at NY 618044: 54.4346, -2.5900] *turn left onto the B6260 Appleby road towards Orton. Go past the M6 motorway interchange, continuing on the B6260 towards Orton.*

The road crosses the upper River Lune, here a strike stream more or less following the basal Carboniferous unconformity (see Chapter 1). The Lune here has an alluvial channel in a wide modern floodplain. You cross the unconformity between the Silurian rocks below and the overlying Carboniferous rocks. Exposed a couple of kilometres to the east of here near Raisgill [NY 636057: 54.4458, -2.5643] are the basal Carboniferous conglomerates overlain by the Carboniferous Limestone sequence of rocks. From the road to Orton there are extensive views back (south)to the northern slopes of the Howgills (Fig. 4.3).

Continue along the B6260 through Orton village. The B6260 bends to the right here, signed for Appleby. Continue along this road for another c.2km. The road begins to climb

4.3 The northern Howgills from south of Orton.

*up the scarp slope of the Carboniferous Limestone escarpment. Just short of the turning for the minor road for Crosby Ravensworth there is an old quarry on the left. Park here. This is **Stop 4.4** Orton Scar* [NY 628098: 54.4827, -2.5764].

Stop 4.4 Orton Scar [NY 628098: 54.4827, -2.5764]

For rockhounds there are exposures of the Carboniferous Limestone here. However, the main purpose of this stop is the view south over the strike-orientated upper Lune valley, with the northern Howgills beyond (see Fig. 1.5). To your east is the scarp slope of the main Carboniferous Limestone escarpment; the rocks dip gently north. In front of you to the south the Lune valley runs east–west, more or less following the base of the Carboniferous rocks. Beyond the Lune valley are the northern Howgills, in Silurian rocks. They are drained by the large northern Howgill valleys visible from here. Bowderdale is to the far left (east), and Langdale and its tributary valleys are to the right (west) of Bowderdale. The even summit is the pre-Pleistocene Howgill summit erosion surface (see Chapter 1).

For those interested either in Carboniferous geology or in limestone geomorphology there is an extra nearby site. Turn left out of the quarry onto the B6260 and continue along the road north-eastwards for c.600m into and through the col, to a second quarry on your right at [NY 633105: 54.4903, -2.5695].

Not only does the quarry face expose the limestone, but just south of the quarry are limestone pavement features.

Return to your car and retrace your route back down the scarp to Orton village. At the south end of the village take the left turn on the B6261 towards Gaisgill. After less than a kilometre (where the B6261 turns sharply right (south) [at NY 629080: 54.4660, -2.5738], *you continue eastwards on a minor road for another 2km to Raisbeck* [NY 644075: 54.4613, -2.5488]. *En route, if you are interested, there is a Neolithic(?) stone circle, a little north of the road* [at NY 640082: 54.4686, -2.5571]. *A track leads to it from the road. From Raisbeck there are two alternative routes. Alternative 1 enables you to see more of the limestone country, including Sunbiggin Tarn and the intermittent Tarn Syke stream. Alternative 2 is the direct route into the upper Lune valley.*

Alternative route 1: via Sunbiggin Tarn. Just beyond Raisbeck [NY 644075: 54.4613, -2.4028] *take the left fork (another minor road) towards Sunbiggin Tarn.*

The Carboniferous Limestone in this area forms a double escarpment including a lower, smaller escarpment formed of rocks near the base of the Carboniferous Limestone sequence. That scarp lies just to the north of the Lune valley and is aligned E–W. The main escarpment (visited at Orton Scar, Stop 4.4), lies a few kilometres to the north. Between the two escarpments is a broad open valley, drained by Tarn Syke, within which a variety of karstic features are developed. Tarn Syke itself may have been originated by glacial meltwater. Nowadays it is an intermittent stream, which under normal flow conditions disappears into sink holes.

Continue from Raisbeck towards Sunbiggin Tarn, crossing Tarn Syke twice (note the tortuous meanders in the tiny channel!). The road comes to a point above Sunbiggin Tarn [NY 675078: 54.4650, -2.5021] – *This is* **Stop 4.5**.

Stop 4.5 Sunbiggin Tarn [NY 675078: 54.4650, -2.5021]

From here there is an overview of the tarn (Fig. 4.4). The tarn itself is probably within a large solutional collapsed depression.

Continue NE for about 2km beyond Sunbiggin Tarn to a junction with a minor road on the right. It makes sense to continue north-eastwards for another 250m or so, to just through the col [at NY 690096: 54.4783, -2.4775]. *To the south of the col are good, clear* limestone pavement features.

Retrace your route back to the previous road junction. From the road junction [at NY 685092: 54.4766, -2.4869], *200m southwest of the col, take the minor road towards the southeast towards Newbiggin-on-Lune and Ravenstonedale.*

It is interesting to note that en route you cross the headwater of Helm Beck, which rises south of the escarpment across a very low divide from the Sunbiggin valley. Helm Beck flows north through a gorge through the escarpment and ultimately into the Eden drainage. This drainage line might have been an early line taken by one of the northern Howgill streams, before capture by the subsequent upper Lune (see Chapter 1).

Continue along the minor road to the junction with the A685 at Newbiggin-on-Lune. Here you rejoin the shorter alternative route 2.

4.4 Sunbiggin Tarn, within the glacial meltwater-incised valley of Rais Beck. The Lower Carboniferous Limestone forms the low ridge in the middle distance; the Howgills form the skyline.

Alternative route 2 (The shorter route): from Raisbeck via Kelleth to Newbiggin-on-Lune.
From Raisbeck continue southeast, across the lower limestone escarpment into the Lune valley, where you join the old A685, near Kelleth. The modern A685 now passes mostly south of the River Lune to Newbiggin-on-Lune, following the line of the old railway. The river has also been diverted along some of this stretch.

En route along the old A685 through Kelleth, you are above the upper Lune valley, with good views beyond Kelleth [at NY 675055: 54.4446, -2. 5024] of the modern meandering channel, locally rapidly migrating, even braided (Fig. 4.5a), and beyond it to the north-eastern Howgills (Fig. 4.5b).

4.5 The upper Lune valley near Kelleth: a) Looking west (downstream); here the river undercuts glacial boulder clay, below the solifluction surface. Note the braided channel. b) Looking south across the upper Lune valley towards Bowderdale in the Howgills; note the solifluction surface in the middle distance and the low terrace just across the river valley. Note also the migrating/braided channels.

Both alternative routes meet at Newbiggin [NY 703053: **54.4354, -2.2751**]. *From here continue on the A685 for less than 2km to a right turn into Ravenstonedale village* [at NY 715046: **54.4313, -2.4330**].

Ravenstonedale lies on Scandal Beck, one of the Eden headwaters. It drains a tiny portion of the NE corner of the Howgills. This is the only part of the Howgills not ultimately within the Lune drainage. Scandal Beck continues NE from Ravenstonedale, through a gorge through the limestone similar to that of Holm Beck, and northwards into the Eden Valley (see also Chapter 1).

Continue south through Ravenstonedale village, then ESE for about 2.5km (lower slopes of the Howgills to your right), to join the A683 (Kirkby Stephen to Sedbergh road). Turn right at the junction [NY 739024: **54.4163, -2.4028**], *towards Sedbergh. After less than 2km* [at NY 734007: **54.4012, -2.4106**] *there is a minor road bearing off to the left which gives better views than the main road. Take this road for about 1.5km to just beyond farm buildings on the right. This is* **Stop 4.6** [SD 730996: **54.3897, -2.4190**]. *Just beyond the farm buildings there is a field entry on the right where you can park.*

Stop 4.6 The Dent Fault [SD 730996: 54.3897, -2.4190]

For the last few kilometres, more or less since joining the A683 after Ravenstonedale, you have been following the outcrop of the Dent Fault, a major regional structure that separates the folded Lower Palaeozoic rocks of the Lake District (including the Howgills) from the near-horizontal Carboniferous rocks of the Pennines (Fig. 4.6a; see also Fig. 1.4). From this viewpoint the Howgills (Harter Fell) lie across the fault to your west. Note the large linear gullies on the steep eastern face of Harter Fell (Fig. 4.6b). South of Harter Fell there are fold structures in the lower Palaeozoic rocks that bring Ordovician rocks (Ashgillian shales and the thin inter-bedded volcanics) to the surface in the cores of a series of anticlines (see Chapter 1). To the east of the fault the Carboniferous Limestone lies adjacent to the fault, but is overlain by the Yoredale series of rocks (alternating sandstones, shales and limestones) that form the flanks of Wild Boar Fell further to your east. The fault itself, a deep-seated reverse fault with some lateral expression, is one of the three major deep-seated regional structures (The Craven, Dent and Pennine faults) that separate the relatively simple Pennine structures to the east from the more complex 'Lake District' structures to the west. The fault itself here more or less parallels this minor road (Fig. 4.6a) a little to the west. A little to the east of the fault, but to the west of the road, are lines of solutional and collapse features in the Carboniferous Limestone.

Continue SW along the minor road to where it rejoins the A683 at Rawthey Bridge [at SD 715979: **54.3653, -2.4353**]. *Note that since just before the last stop you have been back into southward-flowing Rawthey/Lune drainage. Turn left onto the A683; continue for about 1.5km to just short of the Cautley Cross-Keys Temperance Inn, where there is a car park on the right. Park here; this is* **Stop 4.7** [SD 697969: **54.3670, -2.4660**]. *This will also be the parking place for Stop 1 on the southern Howgills excursion (see Chapter 7).*

(a)

(b)

4.6 The Dent Fault: a) Looking NNE along the alignment of the Dent Fault that separates the Howgill terrain on the folded Lower Palaeozoic rocks of Harter Fell (to the left, west) from the Pennine terrain on gently easterly dipping Carboniferous rocks (to the right, east). The fault itself is parallel with and a little to the left (WNW) of the wall. b) Harter Fell across the Dent Fault; note the linear gully systems.

Stop 4.7 Cautley [SD 697969: 54.3670, -2.4660]

Walk about 250m south along the road for an overview of the Cautley cirque area (note that a detailed field visit is recommended in Chapter 7, where a detailed description is given). From this viewpoint the main features are visible (see Fig. 4.7). In the foreground is the alluvial channel of the River Rawthey, where it makes a large bend towards the west. You have been following the River Rawthey since you rejoined the main road at Rawthey Bridge. From there until just upstream of Cautley it has an incised bedrock channel. Beyond the river is the suite of landforms comprising the Cautley cirque area. Behind, to the left are the cliffs of the backwall of the cirque itself (Cautley Crags). To the right is the waterfall of Cautley Spout (supposedly the highest composite fall in England). Above and to the right is the col into the head of Bowderdale (see also Fig. 1.6). The effects of the southward glacial diversion of the drainage have already been described in Chapter 1. Cautley Holme Beck, rising from a valley behind the cirque, used to flow through the col into the head of Bowderdale, but glacial erosion of the cirque diverted it into its present course into the Rawthey. In the centre of the view in front of you (Fig. 4.7) is a mound of uncertain origin, perhaps a former moraine

4.7 Cautley: looking west into the Cautley cirque area from south of the Cautley Cross Keys Temperance Inn. Cautley Crags, at the back of the cirque to the upper left. The original drainage flowed from behind the ridge beyond the crags, northwards through the col into Bowderdale and was diverted glacially over The Spout waterfall into the Cautley valley. In the middle distance the mound below the scree slopes (to the right) may be morainic, but has been interpreted as glacially overrun slope failure (see Chapter 7). Note also the sloping solifluction surface in the foreground, just beyond the River Rawthey.

or a rockfall deposit. (see Chapter 7). In front of you and forming the gentle hillslope to the right is a solifluction surface, produced by solifluction activity during the Late Pleistocene after the area became ice free (see Chapter 2).

Return to your car; rejoin the A683, turning right out of the car park and head for Sedbergh.

Note that this part of the route also features in the southern Howgills excursion (see Chapter 7), so only very brief notes are given here. The road follows the Rawthey valley. For the central third of the way the Rawthey is incised into bedrock. Steep Howgill catchments feed the Rawthey, Hobdale Beck, Ashbeck, and behind Sedbergh, Settlebeck Gill (see Chapter 7).

The circum-Howgill reconnaissance excursion ends in Sedbergh. If your journey originated in Kendal, take the A684 from Sedbergh to Kendal.

CHAPTER 5

The Western Howgills – Carlingill

Introduction

The Carlingill valley has been a major focus of geomorphological research in the Howgills for 40 years or so, primarily by the author, but often in collaboration with colleagues from the University of Liverpool and elsewhere. The research has had two main foci: modern processes (see Chapter 3), and the Late Pleistocene to Holocene landform evolution (see Chapter 2). The highlights of this excursion are the modern gully systems, the Holocene sequence of erosion and deposition, especially in relation to the evolution of the valley floor, and at the head of the valley the glacial (?) diversion of the Carlingill headwaters through the spectacular Carlingill gorge. Allocate most of a day in the field for the full field trip, involving a hike of c.7+km. Note that the last 2km to the end of the hike and back are over steep, rough terrain. If that part were omitted you should allow maybe a half to two-thirds of a day. For the location of the route within the Carlingill valley see the Google Earth image (Fig. 5.1).

Highlights: Modern gully erosion processes. The Holocene sequences of gully erosion and stream deposition. The glacial(?) diversion of the Carlingill headwaters through spectacular Carlingill gorge. The Alluvial fan depositional sequence at Blakethwaite.

The field sites – Carlingill

Start: Park on Howgill Lane at Carlingill bridge on its western side **[SD 624996: 54.3900, -2.5801]**: *space for one car only, without obstructing the animal gate. Alternatively park on the grass shoulder on Gibbet Hill, along the lane c.200m to the south of the bridge* **[SD:624995: 54.3900, -2.5800]**, *where there is space for perhaps three cars. Proceed on foot from Gibbet Hill eastwards, maintaining your elevation above the valley for about 300m to* **Stop 5.1** *(Grains Gill viewpoint:* **[SD 627997: 54.3913, -2.5751]***).*

Stop 5.1 Grains Gill viewpoint [SD 627997: 54.3913, -2.5751]

If you were to only visit one field site in the Howgills, this should be it! The view in front of you (To the north) encompasses much of what is the core of the Late Pleistocene to modern geomorphology of the Howgills (see Frontispiece and Fig. 5.2). The skyline in front of you is formed by a pre-Pleistocene erosion surface (see Chapter 1). Set below the erosion surface, beyond the col at the head of Grains Gill is Archer Moss, site of Cundill's (1976) Holocene pollen diagram (see Chapter 2). Cut into the erosion surface is the Pleistocene valley of Grains Gill with its tributary from the East, Weasel Gill, both tributary to the main stream of Carlingill below you. The smooth surface forming the divide between Grains and Weasel Gills is the Late

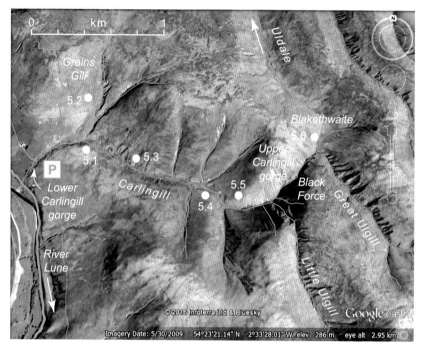

5.1 Google Earth image of the Carlingill valley, showing major excursion stop locations (5.1–5.6).

Pleistocene solifluction surface on soliflucted glacial till (see Chapter 2). Cut below that surface are the Holocene valleys of Grains and Weasel Gills. Both valleys are flanked by now-vegetated gully systems. These, together with the larger incised tributary gully joining Grains Gill from the west, are probably remnants of the tenth-century AD gully systems (see Chapter 2). These gully systems fed sediment into the valley, forming a relatively confined alluvial fan towards the confluence of Grains Gill with Carlingill (see Frontispiece and Fig. 5.2). This is the large fan in front of you that forms a low terrace-like feature where it has been dissected by Carlingill. The lower part of a peat body buried by this fan was radiocarbon dated to 2290 +/-80 BP (Cundill, 1976), making the overlying fan sediments considerably younger, perhaps the same age as the widespread tenth-century AD fan deposits. To the left (west) of the fan is the late Pleistocene main terrace of Carlingill, through which the modern channel of Carlingill cuts into bedrock. At this point the modern channel underwent lateral erosion into the fan toe during the December 2015 flood (see Chapter 3). To the west, but not really visible from this point, Carlingill cuts a steep bedrock gorge down to the River Lune below. On the right of Grains Gill (to the East thereof) about 500m in front of you is the Grains Gill gully system (Fig. 5.2), the site of much of my research work on gully erosion maintained for over 30 years at various scales (Harvey, 1974, 1977, 1987a, 1987b, 1992, 1994, 2000, 2001; see also Chapter 3). This is the next stop (**5.2**) on this excursion.

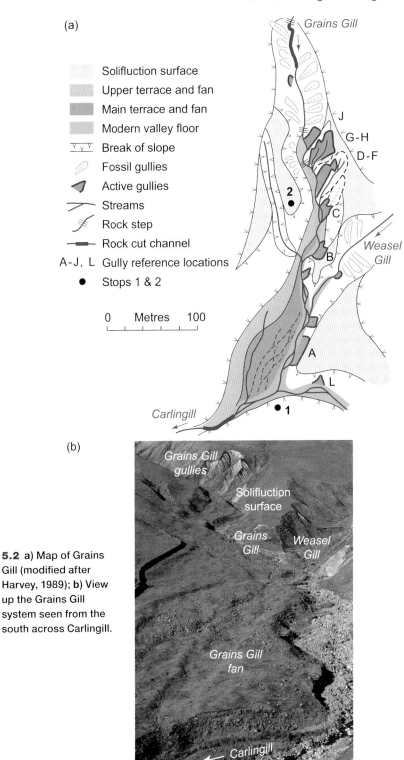

5.2 a) Map of Grains Gill (modified after Harvey, 1989); **b)** View up the Grains Gill system seen from the south across Carlingill.

*Transfer to **Stop 5.2**: Proceed northeast: there is a rough track down into the Carlingill valley. Cross Carlingill, using stepping stones (The water is usually fairly shallow) and walk up the Grains Gill fan, keeping to the west of Grains Gill channel. Immediately on your right is an erosional scar (Gully A on Fig. 5.2a), cut through the solifluction surface into the undisturbed glacial till below. Continue up the fan beyond the Weasel Gill confluence and around a bend to the right (past another erosional scar on the right: Gully B on Fig 5.2a). It is worth clambering up the slope on your left a little to about* **[NY 628003: 54.3953, -2.5741]**, *to obtain an overview of the Grains Gill gully site. This is **Stop 5.2***.

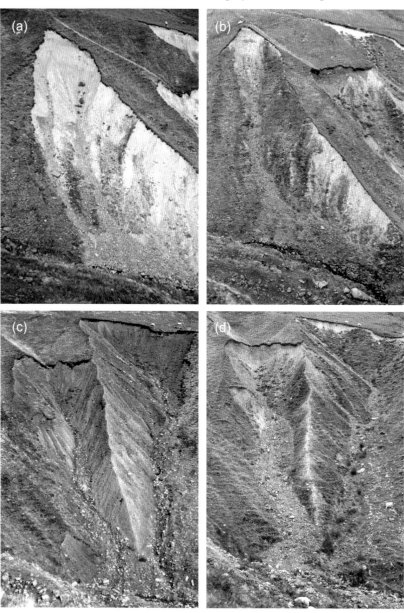

5.3 Sequential photographs showing progressive changes in gully morphology: **a,b)** Gully J 1969, 1999; **c,d)** Gullies G–H 1974, 1997 (for specific gully locations see Fig. 3.4).

Stop 5.2 The Grains Gill gully site [NY 628003: 54.3953, -2.5741], (see also Figs 3.3, 3.4)

The main results of the monitoring of Grains Gill gully site were presented in Chapter 3, relating to erosion rates and seasonality of processes, which in turn are reflected in the seasonality of the surface morphology (Figs. 3.5, 3.6). On the basis of the longer-term development, involving headwards erosion and progressive surface stabilization by base-up revegetation, a simple model of gully development was presented (Fig. 3.7). The types of decoupling sequence involved can be illustrated here by the view in front of you and by Figure 5.3, that shows progressive change in surface morphology between 1969 and 1999 (on Gully J: Fig. 5.3ab) and 1974 and 1997 (on Gullies G/H: Fig. 5.3c,d). The 1969–1989 progressive change in the status of the eroding gully slopes is illustrated in Figure 5.4. The relatively minor, more recent, changes brought about by the December 2015 storm (some renewed erosion on Gully H, though not shown in Fig. 5.4) are also apparent from where you are standing.

Transfer to **Stop 5.3**: Middle Carlingill valley. Retrace your steps back to the Grains Gill/Carlingill confluence. It is best to cross Carlingill there, turn left, and walk upstream,

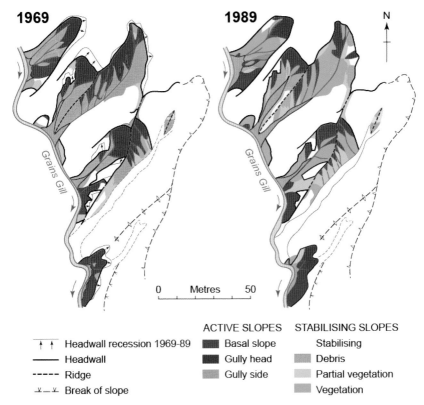

5.4 Grains Gill gullies: map showing progressive change in gully status 1969–1989 (modified from Harvey, 1992)

recrossing the stream as appropriate, for example near the first basally eroded gully/scar (Gully L on Fig 5.2a) on the north bank of the stream. There is a series of points of interest along the valley, all referred to here as **Stop 5.3**.

Stop 5.3 Along Carlingill valley – other gully sites (from [SD 627997: 54.3915, -2.5735] to about [SD 635994: 54.3895, -2.5678])

For location see Google Earth image (Fig. 5.1) and map (Fig. 5.5). As you walk up the valley floor notice the following features. By and large the steep south side of the valley is cut into bedrock at the base, but the north side (That most prone to gullying) is in soliflucted glacial till. Just upstream of the Grains Gill confluence there is a modern basally eroded gully/scar on the north bank of the stream (Fig. 5.6a; Gully L of Harvey, 1992, 1994). A little further on a much larger, but a still basally coupled modern gully occurs, again on the north bank of the stream (Fig. 5.6b; Gully M of Harvey, 1992, 1994). Both gullies underwent basal erosion during the December 2015 flood. Basal erosion on Gully M generated a massive debris flow that blocked the stream, causing it to migrate south across the floodplain (compare Fig. 3.20a, after the flood, with Fig. 5.6b prior to the flood).

Further on still is a large modern but now decoupled and revegetating gully, again on the north bank of the stream (Fig. 5.6c; Gully U of Harvey, 1992, 1994). These three gullies clearly illustrate gully coupling characteristics (Harvey, 1997) and the applicability of the gully development model presented earlier (Fig. 3.7). Significantly, Gully U (not coupled to the basal stream) did

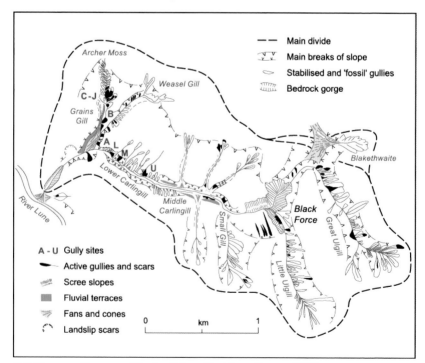

5.5 Geomorphic map of Carlingill (modified from Harvey, 1992).

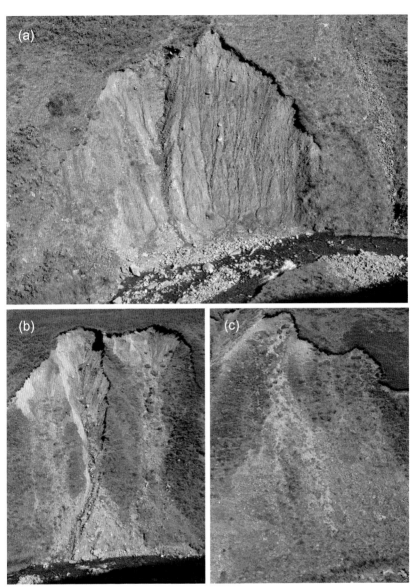

5.6 Active gullies in the Carlingill valley; photos taken in 2012. a) Gully L; b) Gully M; c) Gully U; (for locations see Fig. 5.5). Compare Fig. 5.6b (Gully M: pre-2015 flood) with the site after the 2015 flood (see Fig. 3.20a).

not undergo erosion during the December 2015 flood, although there was a shallow slope failure on the open slope to its north, which fed sediment to the valley floor (see Chapter 3). Initially that sediment was incised by the stream, but has since become decoupled and started to stabilize. In 1998 on the hillslope to the left (west) of this gully there was also a small hillside slope failure that fed debris-flow lobes down the hillslope, one of which reached the valley floor as a small debris-flow frontal lobe immediately to the west of Gully U (Harvey, 2001).

Once beyond the bedrock channel section immediately upstream of the Grains Gill confluence, the channel is an alluvial channel occupying the valley floor, on which there is evidence of earlier abandoned channels (discussed below in relation to Stop 5.4). Prior to the December 2015 flood, through most of this reach of Carlingill, the stream channel was a single thread channel with widths of the order of 5–10m. However, in zones of high coarse sediment input, at tributary junctions or below eroding gullies, the channel locally flipped to a braided regime with total widths of over 20m (see Fig. 5.7) (Harvey, 1997).

You now come to where the valley floor widens in the vicinity of a tributary junction from the north, Haskaw Gill. This is **Stop 5.4**: *Middle Carlingill.*

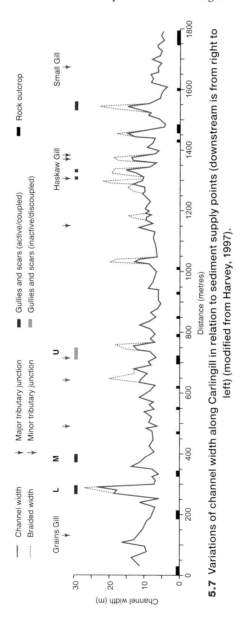

5.7 Variations of channel width along Carlingill in relation to sediment supply points (downstream is from right to left) (modified from Harvey, 1997).

Stop 5.4 Middle Carlingill: Holocene and recent valley floor evolution [SD 635994 to SD 640993: 54.3888, -2.5629 to 54.3871, -2.5568] For location see Google Earth image and Carlingill map: (Figs 5.1, 5.5).

In this part of the valley the late Holocene valley-floor landforms are exceptionally well developed (see Fig. 2.9) and have been studied and dated using lichenometry. Figure 3.9a shows the sedimentological/geomorphic features of this part of the valley floor (surveyed in 1984), together with the lichen zones (Fig. 3.9b – calibrated as on Fig. 2.3). From this mapping we have interpreted the sequence of channel changes and bar deposition, involving channel migration, cutoff and avulsion (Fig. 5.8). The valley floor

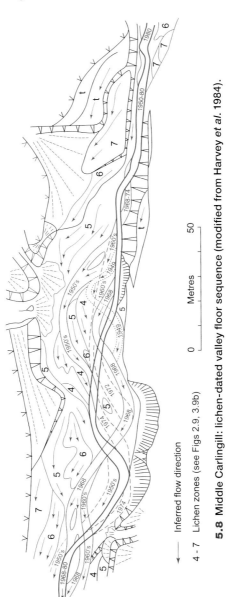

5.8 Middle Carlingill: lichen-dated valley floor sequence (modified from Harvey et al. 1984).

Inferred flow direction

4 - 7 Lichen zones (see Figs 2.9, 3.9b)

0   50
   Metres

landforms also exhibit a clear soil chronosequence (see Chapter 2; see also
Harvey *et al.*, 1984). Figure 2.2 illustrates the sequence on the older surfaces.
Figure 5.9 shows soil characteristics, profile, pH and organic content on
younger representative lichen-dated sites on the valley floor (after Harvey

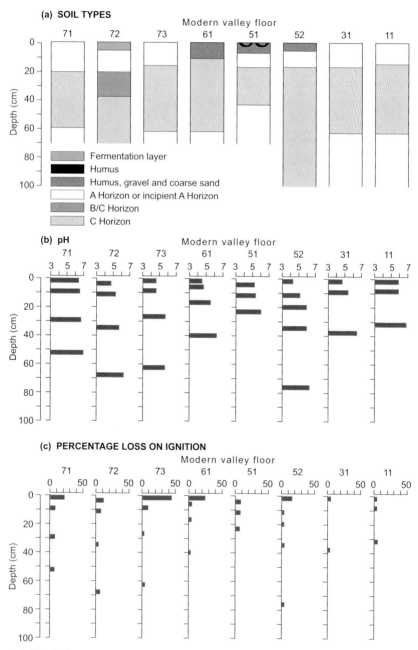

5.9 Middle Carlingill: soil profile details for soils developed on the valley floor: profile
numbering relates to the lichen zones identified on Fig. 3.9b (modified from Harvey
et al., 1984).

et al., 1984). This reach also underwent channel change during the December 2015 flood, by bank erosion and channel widening, with an area at the downstream end of this reach flipping to a braided regime.

The alluvial valley floor persists a little further upstream to just beyond the Small Gill confluence, to the outlet of the gorge of upper Carlingill [SD 640993: 54.3871, -2.5568] (Fig. 5.10). The view from here up the valley of Small Gill includes major now-stabilized tenth-century gully systems (see Chapter 2), especially on the east side of the valley.

It is at this point that you must decide whether to continue into the upper part of the Carlingill valley. It is spectacular, and well worth the effort, but involves scrambling up the rubbly slope to your north and traversing the steep slope on a narrow path above the main body of scree. If you decide to return from here to your car, simply follow Carlingill back downstream to Carlingill Bridge.

*If you decide to continue to **Stops 5.5 and 5.6**, take care! – this is hazardous terrain. From the confluence of Small Gill with Carlingill scramble up the slope towards the northeast. Make sure you climb sufficiently high to avoid the worst of the screes on the lower slopes, and the bedrock outcrops in the upper part of the gorge. There is a crude path that runs east high above the gorge – again, take care! Midway along this path (about [SD 644994: 54.3881, -2.5512]) you come to a spectacular view across the gorge, into the tributary valley of Little Ulgill Beck, and its course into Carlingill over Black Force waterfall. This is **Stop 5.5**.*

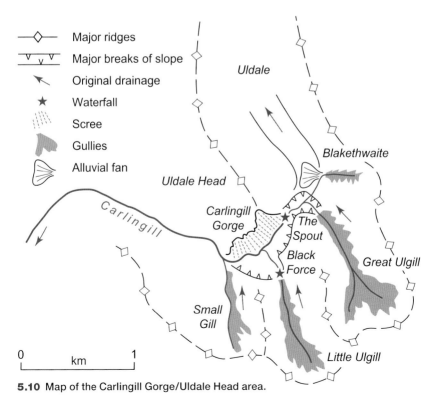

5.10 Map of the Carlingill Gorge/Uldale Head area.

Stop 5.5 Viewpoint over Carlingill Gorge and Black Force water-fall into Little Ulgill valley [SD 644994: 54.3881, -2.5512]

This viewpoint gives an overview of the area at the head of Carlingill (Figs 5.10, 5.11a,b). At some stage pre-Last Glacial, the present headwaters of Carlingill, including perhaps Small Gill (in front of you to your right), but

5.11 Carlingill Gorge: **a)** General view looking East into Carlingill Gorge; **b)** Black Force waterfall, where Little Ulgill Beck cascades into Carlingill Gorge. Note: A view of the gully systems within Great Ulgill valley is shown in Figure 2.5.

definitely including Little Ulgill (directly in front of you) and Great Ulgill (in front of you to your left), were all headwaters of Uldale. Uldale is a tributary of the Langdale system that drains north to the Upper Lune (see Chapter 1). The former headstreams of Uldale (listed above) have been captured by or diverted into the Carlingill system that flows west to the middle Lune downstream from the Lune Gorge (see Chapter 1). I think it highly unlikely that this is a simple capture induced by the lower base level of the middle Lune. Upstream of its confluence with the Lune, Carlingill flows through another deeply incised gorge. That gorge is downstream of Carlingill Bridge (To the west of where this excursion started). It could obviously have been Lune base-level induced. Quite clearly the effects of that incision have not propagated far up Carlingill. In my opinion, a much more likely explanation of the capture/diversion of the former Uldale headwaters into Carlingill would be glacial or fluvio-glacial diversion at the close of regional glaciation. There are other not dissimilar north-to-south diversions elsewhere in the area (see Chapter 1).

Figure 5.11a shows the view in front of you, of the upper Carlingill gorge. The gorge is fed by Little Ulgill Beck, which has a relatively low-gradient upper valley then cascades into the gorge over Black Force waterfall (Fig. 5.11b). The situation of Great Ulgill Beck is not dissimilar. It has a low-gradient upper valley, then joins upper Carlingill Beck. That stream then enters the gorge by way of a deep incision and another waterfall (The Spout; see Fig. 5.10). Other impressive landscape features here are the deep and complex, now mostly stable, hillslope gully systems, probably of tenth-century origin, in Small Gill, Little and Great Ulgill valleys (see Fig. 2.5) which, as elsewhere, cut into hillslope soliflucted glacial till.

From this viewpoint the rough path rounds the upper part of the gorge and descends into the col at Blakethwaite. This is **Stop 5.6** [SD 648998: 54.3929, -2.5440].

Stop 5.6 Blakethwaite [SD 648998: 54.3929, -2.5440]

At Blakethwaite there is a low-angle alluvial fan (Figs 5.12, 5.13) set into the col created by the drainage diversion of the former Uldale headwaters from Uldale into Carlingill (see above). The fan is fed by now largely stabilized gully systems. The descent into the col provides an excellent overview (Fig. 2.10a). Currently the fan is right on the divide; its northern flank feeds north into Uldale, but its southern flank feeds south into Carlingill. In mid-fan there is a shallow fanhead trench cut into the fan surface, through the fan deposits into the underlying material (Fig. 2.10b). The trench exposes a sedimentary sequence comprising a basal exposure of hillslope soliflucted till, overlain by a peat layer, followed by the fan deposits (Fig. 5.13). We have a series of radiocarbon dates from this section (Harvey, 1996). We have a basal date from charcoal fragments that predate the main peat formation (2975 +/-85 BP), then several dates from the peat itself, but most importantly a date of 1210 +/- 50 BP from the top of the peat, prior to burial by fan deposits. Again, in accordance with a number of other dates from

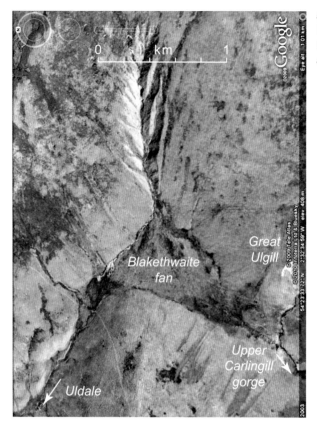

5.12 Google Earth Image of the Uldale Head/Blakethwaite area.

similar sites within the Howgills (see Chapter 2), the indication is of tenth-century (Norse) initiation of hillslope gullying. The gullies fed sediment to the alluvial fan, probably following human-induced vegetation change, rather than as the result of a climatic cause (Chiverrell *et al.*, 2007, 2008). Interestingly, we also have two dates from the soil formed on the fan deposits, the oldest of which (from the humin fraction of the soil) is 890 +/-50 BP, suggesting that by then sedimentation had ceased and soil formation was underway. The name 'Blakethwaite' itself suggest human disturbance: 'thwaite' is a Norse-derived word meaning a woodland clearing. There was a final pulse of sedimentation (at 410 +/-50 BP), burying the thin soil that had formed over the main fan deposits. After this the site recovered, much later to be trenched by the modern fan channel.

The Blakethwaite site forms the end of the excursion. Return to your vehicles back down Carlingill valley the way you came.

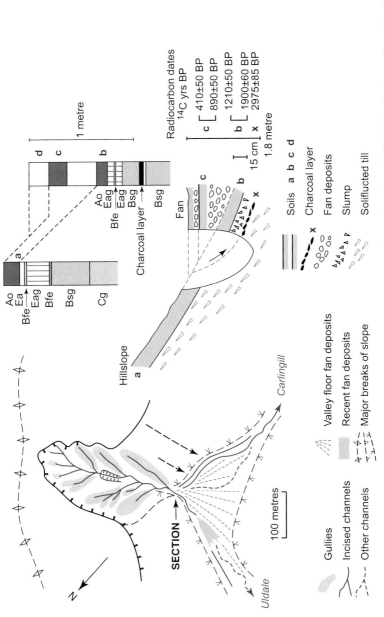

5.13 Blakethwaite. On the left: sketch map of Blakethwaite fan; on the right: sketch section through Blakethwaite fan soils and sediments (modified from Harvey, 1996). For ground photos of Blakethwaite fan and section cut into the fan, see Fig. 2.10 a,b respectively.

CHAPTER 6

The Northern Howgills

Introduction

This Chapter presents two field trips to the largest and most complex of the northern Howgill valleys, Langdale and Bowderdale (Fig. 6.1), with brief mention of a third valley, Weasdale. As in the previous chapter, *locations, instructions, etc. will be given in italic script.* Again, the same word of warning. There are no roads, so after parking, all field visits are on foot. As before the country is rough, sometimes with steep or treacherous slopes. Hiking boots are essential; again, raingear is a sensible precaution! As in Carlingill (see Chapter 5), there is a need to cross streams. If stream discharges are high, modify your route accordingly. Approximate distances are given in kilometres; locations are given by National Grid references and in GPS coordinates.

Highlights: The late Holocene hillslope erosional sequence: gully development, debris-cone and alluvial-fan deposition, ascribed to the Viking period by radiocarbon dating. The modern geomorphic system: channel morphology, modern debris-cone and alluvial-fan deposition, especially following the 100-year storm and flood in 1982; channel response to the December 2015 flood.

A Langdale

Be warned – this field visit requires a full day and involves a long round-trip hike of 16–18km. Although it is mostly on the valley floor and need not involve any major climbs, the terrain is rough. The route is shown on the Google Earth image (Fig. 6.1).

Turn south off the A685 about 3km east of Tebay onto the minor road signed Long-dale (Langdale; both versions seem to be used). After about 500m, park in the hamlet of Longdale **[NY 645051: 54.4397, -2.5493]**. *Take the track that leads due south from Longdale hamlet. Just beyond the farm turn right through a gate (close the gate after you!) into walled pasture fields beside Langdale Beck. Follow these fields southwards alongside Langdale Beck for a little more than 1km, crossing the walls either through gaps or over stiles, into scattered oak trees, on the east flank of the valley floor. Follow the east bank of the stream for about another 1km to the confluence of the Uldale/Churngill tributary (from the west) with the main Langdale stream. Uldale Beck is the stream that flows from the Blakethwaite col at the head of Carlingill, which lost its headwaters by glacial diversion (?) to Carlingill (see Chapter 5 and Figs 5.11, 5.12, 5.13).*

You are now in open country. Continue to follow Langdale Beck upstream, south-eastwards for another c.1.5km. Note the Nevygill confluence, from the south (Fig. 6.1), with its more or less intact stable low-angle tributary-junction alluvial fan, probably a tenth-century feature. Continue to follow Langdale Beck for another c.300m to streamside sections exposed at the foot of three now stable debris cones emanating from

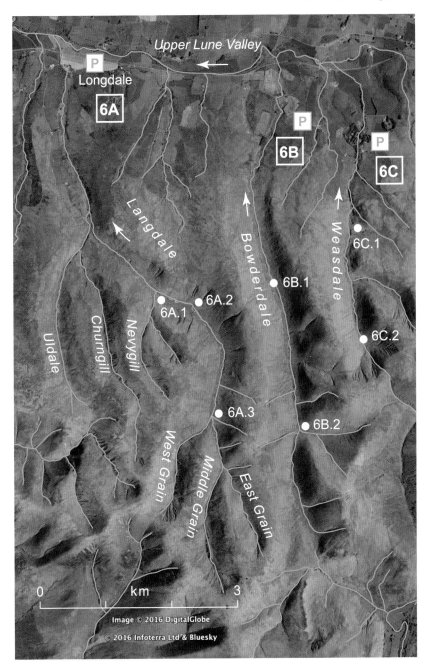

6.1 Google Earth image of the main northern Howgill valleys: Langdale, Bowderdale, Weasdale.

*older, now stabilized gully systems on the north side of the valley. This is **Stop 6A.1**, Burnt Gill, Middle Langdale* **[NY 662016: 54.4079, -2.5204]**(Fig. 6.2; *see also Fig. 2.8).*

Throughout this section of the valley, note the effects of the December 2015 flood (see Fig. 3.20b). In many places lateral erosion by the stream of

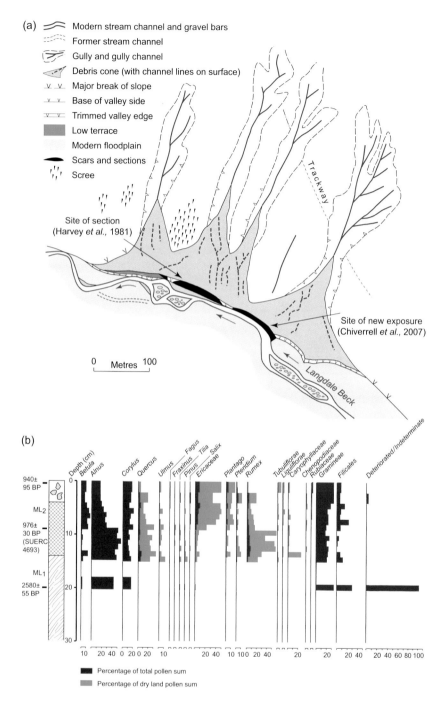

6.2 Middle Langdale (Burnt Gill): **a)** Map of the Burnt Gill debris cones (for photos see Fig. 2.8) showing location of sections (modified from Harvey *et al.*, 1981, Chiverrell *et al.*, 2007); **b)** Pollen diagram and Radiocarbon dates taken from the buried soil (after Chiverrell *et al.*, 2007). For English species names see caption to Figure 2.11.

either floodplain or low terrace sediments caused significant lateral migration and channel widening. In extreme cases this led to the development of new braided reaches. Stream widths now no longer conform to the model presented in Chapter 3 for single thread channels (see Fig. 3.10), but to the cluster of points on the graph well above the general line.

Stop 6A.1 Middle Langdale (Burnt Gill) [NY 662016: 54.4081, -2.5212]

At this site (Burnt Gill) four large now-stabilized hillslope gullies fed sediment into large streamside coalescent debris cones, also now stabilized. The gullies themselves are cut through late Pleistocene hillslope solifluction deposits into the underlying bedrock (see Figs 2.8, 6.2). Langdale Beck has cut sections in the base of the debris cones, exposing soils buried by the debris-cone sediments (Fig. 6.2). The soils themselves are formed on an underlying stream terrace. Two sets of radiocarbon dates are available from the buried soils, the first set from under the second debris cone from the west (Harvey *et al.*, 1981) gives the following dates (in radiocarbon years). The oldest date from the base of the soil developed on the former stream sediments is 2580 +/-55 BP. The youngest date, from organic material formed immediately prior to burial by the debris cone, is 940 +/-95 BP. From pollen analysis of pollen preserved within the buried soil, we have a picture of the local vegetation at the time of soil formation. We have attempted to distinguish between valley-floor 'wetland' species and the more regional 'dryland' species (see Chapter 2). Initially the local 'wetland' vegetation appears to have been dominated by *Alnus* (alder) and *Corylus* (hazel), with a regional 'dryland' contribution by *Quercus* (oak) and *Rumex* (docks and sorrels). Then near the top of the soil there are dramatic changes. Alder decreases significantly, to be replaced in part by *Betula* (birch). Of the 'dryland' pollen, *Rumex* decreases dramatically and is replaced by *Ericacea* (heathers). These vegetation changes indicate a change in human pressure on the land, but pre-date gully development and debris-cone formation. The latter phenomena occurred during or immediately after the tenth century AD, almost certainly following the Norse introduction of widespread sheep grazing on the fells. Another radiocarbon date (Chiverrell *et al.*, 2007) comes from a more recent exposure of the soil at the base of the most easterly of the debris cones, giving a date of 976 +/-30 BP for the upper part of the buried soil, confirming the overall chronology based on the other dates.

From this part of the valley upstream, the channel of Langdale Beck has exhibited chronic instability, especially in response to and subsequent recovery from the 1982 flood (Fig. 6.3). It is a zone with several major sediment supply points, below which even before 1982 there was some braiding. In 1982 a number of channel zones switched to a braiding regime. Even since then recovery has been incomplete. Prior to that flood most of the channel of Langdale Beck had been characterized by a single-thread, relatively narrow channel that had undergone only minor change since 1948, as is evident from the 1948 air photographs. The effects of the 1982

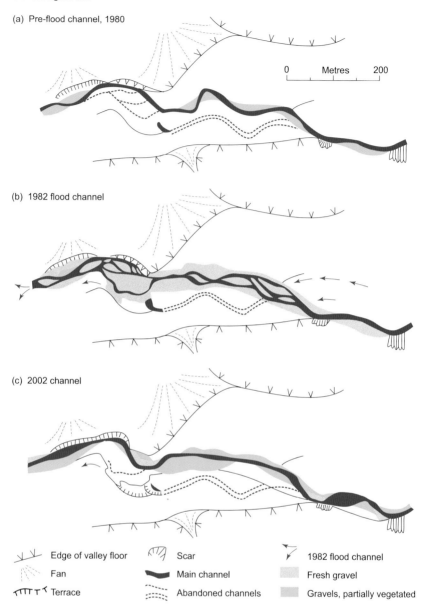

(a) Pre-flood channel, 1980

0   Metres   200

(b) 1982 flood channel

(c) 2002 channel

⌄⌄ Edge of valley floor	⟨⟩ Scar	⌁ 1982 flood channel
Fan	Main channel	Fresh gravel
Terrace	Abandoned channels	Gravels, partially vegetated

6.3 The channel of Langdale Beck in Middle Langdale: **a)** In 1980 (i.e., pre- the 1982 flood); **b)** channel change in response to the 1982 flood; **c)** recovery by 2002 (modified from Harvey, 2007). See also Figure 3.20b for an example of channel morphology created by the 2015 flood.

flood were dominated by bank erosion and channel widening (Fig. 6.3). In places, especially downstream from points of major sediment input, the channel regime switched to braided, with the deposition of extensive gravel bars. This zone was also seriously affected by channel change during the December 2015 flood. During that flood the main sources of sediment were primarily through bank erosion of floodplain and low terrace sediments, rather than through input from tributary gully systems, as had happened during the 1982 flood.

The channel morphology of the Langdale and Bowderdale channels has been studied and two regimes can be identified (see Chapter 3). For the relatively stable single-thread channels, width can be related to drainage area (and so to formative discharge) (Harvey, 1987b) by the regression equations quoted in Chapter 3 (see Fig. 3.10). For the braided reaches, downstream of active sediment supply points (including several pre-flood braided reaches, as well as those that braided as a result of the 1982 flood), an alternative regression was been calculated (see also Chapter 3 and Fig. 3.10).

During the 1982 flood, many hitherto stable single-thread channels in the Langdale and Bowderdale valleys jumped to the braided regime (Harvey, 1987b, 1991) (Fig. 3.17). Since then, many, but not all, of these channel reaches had recovered by reverting to single-thread status (Harvey, 2007) (Figs 6.3, 6.4). This trend has been interrupted by the 2015 flood, with the formation of many new unstable locally braided reaches (see Fig. 3.20b).

Continue along the valley for another 300–400m to a major debris cone issuing from a gully system on the north-east side of the valley. This is Thrush Gill (Figs 3.14; 3.15a; location on Fig. 6.1), **Stop 6A.2** [NY 662012: 54.4044, -2.5149].

Stop 6A.2 Thrush Gill (and neighbouring sites) [NY 662012: 54.4044, -2.5149]

We are now clearly in the part of Langdale strongly affected by the 1982 flood (see Chapter 3; see also Harvey, 1986). Many of the gully systems in both Langdale and Bowderdale (see below; Chapter 6, Section B) were reactivated by the effects of that storm, feeding sediment to debris cones and alluvial fans where the gully systems issue into the main valley, burying the earlier tenth-century cones and fans. Since then, the surfaces of many of the fans/cones have stabilized and revegetated (see Fig. 3.18).

Here and in Bowderdale, 13 debris cones/fans were deposited (Figs 3.12, 6.5). Their sedimentology ranged from cohesive debris flows, through transitional stony deposits to coarse fluvial cobble bars and finer fluvial gravel sheets (see Chapter 3; see also Fig. 6.6). Thrush Gill was one of the major 1982 cones. Its sediments were dominated by the deposition of cohesive debris flows (Figs 3.14, 3.15a). Since then the surface has become partially stabilized by revegetation. Compare Figure 3.15a with the situation on Thrush Gill cone today. When we (Wells and Harvey, 1987) studied these features, we grouped the fans/cones in Langdale and Bowderdale (see also Chapter 6, Section B on Bowderdale) into four categories. This was done on the basis of the dominant facies of their constituent sediments (see Figs 3.12, 3.14, 6.5; Table 3.3), essentially reflecting the water:sediment ratio during deposition. Thrush Gill is at one end of that spectrum, dominated by cohesive debris-flow deposits, and characterized by a lobate surface morphology (see Fig. 3.15a), still evident today.

Within this same section of the valley are two further 1982 flood debris cones (Figs 3.12, 6.5, 6.6), both fed from gully systems on the east side of the

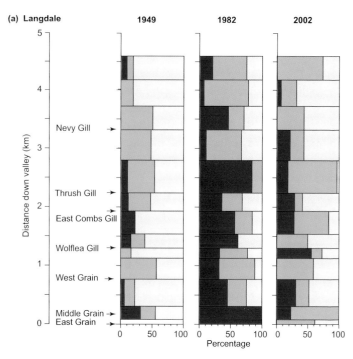

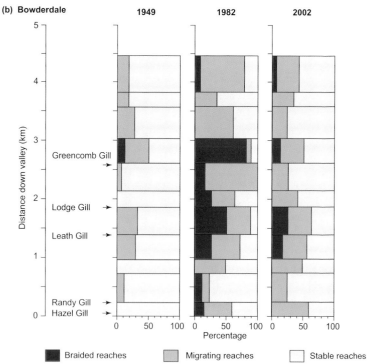

6.4 Langdale and Bowderdale Becks – channel evolution. The proportions of channel segments classified as stable, migrating and braided along the valleys in 1949 (based on air photo analysis), in 1982 (after the flood, based on field survey), and in 2002 (after recovery, based on field survey) (modified from Harvey, 2002).

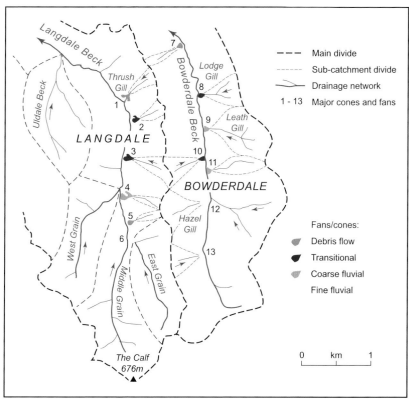

6.5 The distribution of debris cones in Langdale and Bowderdale activated by the 1982 flood, showing their dominant facies. Numbering system as for Figure 3.12 (Langdale: 1. Thrush Gill; 2. East Combs Gill; 3. Wolflea Gill; 4. West Hazel Gill; 5. New Gill; 6. East Grain. Bowderdale: 7. Greencomb Gills; 8. Lodge Gill; 9. Leath Gill; 10.Woofler Gill; 11. Wotey Gill; 12. Hazel Gill; 13. Rams Gill).

valley, East Combs Gill [NY 666012: 54.4015, -2.5122] and Wolflea Gill [NY 667003: 54.3966, -2.5133], both of transitional sedimentology (see Tables 3.3, 3.4). What is evident from all the fans and cones in Langdale and Bowderdale is that the smaller catchments tend to have produced dominantly cohesive debris-flow cones, and the larger catchments tend to have produced dominantly fluvial fans. Also evident is the role of the depositional gradient; lower gradient fans show higher fluidity in their depositional facies than do higher gradient fans and cones (Table 3.3). Position within the storm also seems to have had an effect. The storm appears to have been centred over the upper reaches of Langdale and Bowderdale. Debris-flow dominated fans/cones tend to be further away from that centre, with fluvially dominant fans within the storm centre (Fig. 6.5). There is one major exception: New Gill, near the head of Langdale, is a very small debris-flow dominated cone, fed by a very small steep catchment. A further interesting phenomenon is that even during deposition during the storm, each fan showed evidence of increasing, then decreasing fluidity of the transport process (Fig. 3.16).

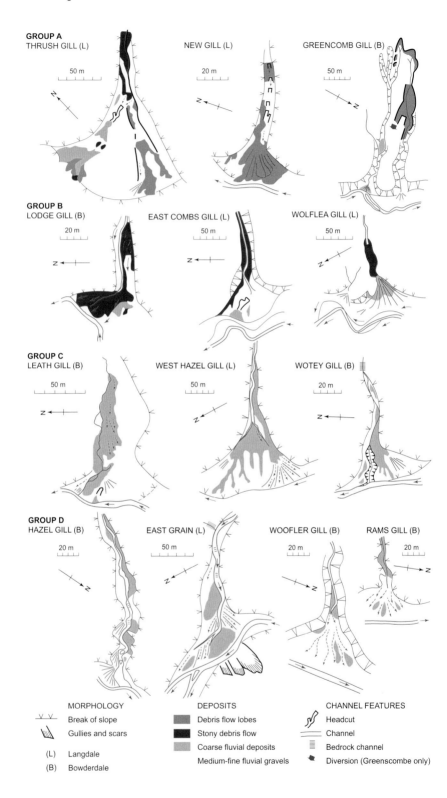

GROUP A
THRUSH GILL (L)
NEW GILL (L)
GREENCOMB GILL (B)

GROUP B
LODGE GILL (B)
EAST COMBS GILL (L)
WOLFLEA GILL (L)

GROUP C
LEATH GILL (B)
WEST HAZEL GILL (L)
WOTEY GILL (B)

GROUP D
HAZEL GILL (B)
EAST GRAIN (L)
WOOFLER GILL (B)
RAMS GILL (B)

MORPHOLOGY
∨ ∨   Break of slope
      Gullies and scars

(L)   Langdale
(B)   Bowderdale

DEPOSITS
      Debris flow lobes
      Stony debris flow
      Coarse fluvial deposits
      Medium-fine fluvial gravels

CHANNEL FEATURES
      Headcut
      Channel
      Bedrock channel
      Diversion (Greenscombe only)

6.6 (Opposite) The 13 debris cones and fans in Langdale and Bowderdale activated by the 1982 flood, grouped into four groups on the basis of the sedimentology of their constituent deposits. Group A dominated by cohesive debris flows; Group B by transitional deposits; Group C by coarse fluvial gravels; and Group D by finer stratified fluvial gravels (modified from Wells and Harvey, 1987). For more detailed maps of the four type examples (Thrush, Lodge, Leath and Hazel Gills) see Figure 3.14.

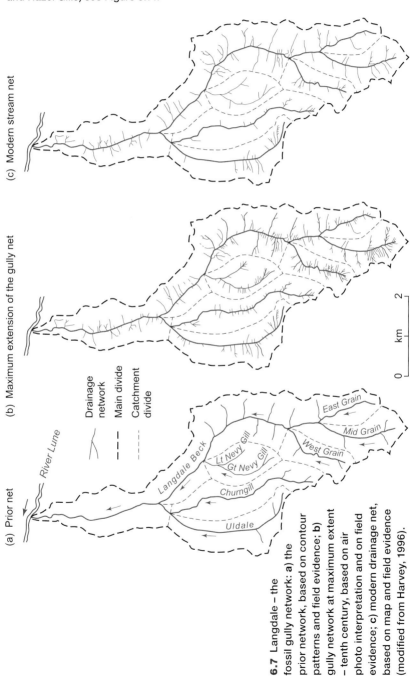

6.7 Langdale – the fossil gully network: **a)** the prior network, based on contour patterns and field evidence; **b)** gully network at maximum extent – tenth century, based on air photo interpretation and on field evidence; **c)** modern drainage net, based on map and field evidence (modified from Harvey, 1996).

Interestingly, none of the 1982 cones and fans showed any significant erosion or deposition in response to the December 2015 flood. On most, the at least partial post-1982 revegetation was little disturbed during the 2015 flood. On some cones, the fan channel showed very minor incision, with perhaps a little deposition of fresh sediment, but as a whole they were not affected by that flood. Nor did the feeder gullied catchments show much erosional change. These characteristics reflect the differences between the two flood events. The 1982 event was sediment-led, but the 2015 event was flood runoff-led, primarily causing basal erosion by channel processes. In 1982 the fans were the zones of greatest geomorphic activity; in 2015 it was the stream channels.

Continue along the valley floor to the confluence of West Grain with Langdale Beck [SD 666997: 54.3920, -2.4922]. *You are now in Upper Langdale: **Stop 6A.3** (see Fig. 6.1).*

Stop 6A.3 Upper Langdale [SD 666997: 54.3920, -2.5154].

All three headwater catchments of the Langdale Beck (West, Middle and East Grains) are deeply dissected by the (presumably) tenth-century AD gully networks (e.g. Figs 6.7, 6.8). In this area there are three 1982 debris cones/fans (Table 3.3; Figs 6.5, 6.6): West Hazel Gill [SD 667996: 54.3924, -2.5154], a coarse, fluvially dominant cone; New Gill [SD 667994: 54.3883, -2.5149], a debris cone from a small steep gully system; and East Grain fan [SD 667991: 54.3858, -2.5129] (Fig. 3.18a), a wholly fluvial fan, fed by a large gullied headwater catchment. During the 1982 storm there were numerous shallow slope failures on the valley sides in this area (Fig. 6.9a). The channel of Langdale Beck between the West and East Grain confluences is confined within a narrow valley floor, but the valley floor underwent radical change during that flood (Fig. 6.9b).

From this area we can see into the headwaters of two of the main feeder channels to Langdale (East and Middle Grains). This headwater area was intensely gullied during the hillslope gullying phase in the tenth century AD. The valley-side gully networks tend to be primarily linear with only minor branching, but the valley-head networks tend to be dendritic with quite complex branching patterns (see Chapter 2; Fig. 2.6). Perhaps the best view of one of these headwater systems is the head of West Grain – see below.

East Grain fan marks the end of the itinerary – to return to your vehicle retrace your steps down the valley.

En route: If you want a good view of a valley-head gully system, the best developed and that most easily seen is probably the head of West Grain. From the West Grain tributary junction, clamber obliquely up the northern slope of the West Grain valley to a point [about SD 663996: 54.3917, -2.5184] in mid-slope, from where the gully system can easily be seen (see Fig. 6.8).

End of the excursion. Return to Langdale village, down the valley, the way you came.

B Bowderdale

Be warned – this field visit, as for Langdale, requires most of a day and involves a fairly long round-trip hike of c.12km (but less than that for Langdale), although it is mostly on

(a)

(b)

6.8 Now-stabilized fossil gully system at the head of West Grain, Langdale: **a)** general view: note contrasting network complexity between valley-side and valley-head gully systems; **b)** detailed view of valley-head gully systems.

the valley floor and need not involve any major climbs. The route is shown on the Google Earth image (Fig. 6.1).

On the A685 about 7km east of Tebay, or c.4km west of Ravenstonedale, turn south onto the minor road towards Bowderdale. After about 200m the road to Bowderdale turns right (to the west), but you should continue south for another c.800m towards Scar Sikes. Where the metalled lane turns left (east) towards Scar Sikes farm, you should park on the rough ground adjacent to the junction **[NY 683043: 54.4322, -2.4922]**. On foot, follow the track S and SW up the spur that separates the Bowderdale valley (to your right – west) from a minor valley (to your left – east). Continue along this track, descending into the Bowderdale valley **[at about NY 674033: 54.4229, -2.5050]**. Follow the eastern side of the valley (there is a crude path along the lower part of the hillslope, a little above the *valley floor) upstream for about 2km into Middle Bowderdale (**Stop 6B.1** – [NY 675017: 54.4107, -2.5008]) at Greencomb Gills. These are two small debris cones fed from gullies on the western hillslope.*

6.9 Effects of 1982 flood at the head of Langdale valley: **a)** small valley-side slope failures, rill development and debris cones; **b)** lateral migration, scour and deposition: the channel of upper Langdale.

On your route through the lower part of the Bowderdale valley, note some of the geomorphic features. The valley sides, especially the western valley side (Fig. 1.7), are blanketed by soliflucted glacial till, forming extensive low-angle sloping surfaces. On the east side of the valley, because of the thinner periglacial mantle downslope from periglacial screes, the thinner, stony periglacial sediments exhibit periglacial patterned ground of stone polygons and garlands (see Fig. 1.8c) . Within the valley itself and cut into the solifluction surface, there is a suite of fluvial terraces, the uppermost of which has been radiocarbon dated to pre-date 6070 +/-70 BP (Miller, 1991), and is therefore probably Latest Pleistocene to early Holocene in depositional age. There are (undated) lower terraces, but the modern valley floor has been dated as post 600 +/-80 BP (Miller, 1991). There are three available radiocarbon dates on older tributary gully/fan relationships in this part of Bowderdale (Dunsford, 1998), including two *c.* tenth-century dates (see below). There is a younger one of 300 +/-40 BP, for the top of a buried soil below a small tributary cone below West Fell, a little further up the valley on its west side. This date is considerably younger than the cluster of tenth-century dates and may be indicative of a younger (post-medieval?) erosional phase.

The modern channel of Bowderdale Beck through Lower Bowderdale is a single-thread gravel-bed wandering channel (Al Farraj and Harvey, 2010). In this lower reach the 1982 flood had relatively little impact compared with reaches further upstream or compared with the channel of Langdale Beck. There was some bank erosion and limited lateral migration, plus the deposition of relatively minor gravel spreads on the valley floor. Downstream from Greencomb Gill there was little direct sediment input into the channel. The effects of the December 2015 flood were similar to those on Langdale Beck, but less dramatic. There are a number of new and re-activated valley-side scars, downstream of which there is clear evidence of local sediment input into the stream. More important was bank erosion of floodplain and low terrace sediments, and associated lateral channel migration, resulting in a number of reaches switching from stable single-thread morphometry to that of a wider, laterally unstable channel regime. This process becomes more important in the middle reaches of Bowderdale Beck (upstream from about [NY 675020: 54.4132, -2.5022] to about [NY 677010: 54.4040, -2.4979]). Otherwise it is too early to assess the quantitative impact of the 2015 flood event in much more detail; Professor Richard Chiverrell (University of Liverpool) is still working on it!

Site 6B.1 Middle Bowderdale

This section of the valley shows a complex network of deep Holocene gully systems on the valley sides (Fig. 6.10), many of which fed tenth-century fans and cones, a number of which have yielded radiocarbon dates and/or have provided Holocene palaeo-environmental information. This part of the valley also includes some of the most significant tributary junction fans/cones within

6.10 General view of Bowderdale looking up-valley. Note the col at the head of the valley between upper Bowderdale and Cautley cirque (see Chapters 4 and 7). Note the valley-side gully systems in middle Bowderdale, and the modern valley floor incised into solifluction surfaces, within which the modern channel is locally migrating and locally braided.

the Howgills, which were affected by the 1982 flood (Table 3.3). At the same time, major channel changes occurred on Bowderdale Beck by avulsion and braiding involving sedimentation on the valley floor.

At Greencomb Gills [NY 675017: 54.4107, -2.5008] are two small fans/debris cones fed by hillslope gullies from the western valley side. Note that the two cones exhibit differential coupling relationships between the gully/cone system and the main channel of Bowderdale Beck (see Fig. 3.11). Of these, the southernmost gully system feeds sediment directly into Bowderdale Beck, through a now dissected tributary junction fan. This gully is coupled to the Bowderdale fluvial system. The more northerly gully system joins the main valley at a location away from the channel of Bowderdale Beck. Its sediment accumulates in an active fan at the margin of the valley. This gully system is not coupled to the Bowderdale fluvial system.

A little further upstream at Thickcomb Gill [NY 677014: 54.4068, -2.4987] is another west-bank gully system. This cone fed little sediment into Bowderdale in 1982, but it is an important site because the older cone has been radiocarbon dated here by Dunsford (1998). Two dates from the top of the buried soil are 930 +/40 BP and 790 +/40 BP (Dunsford, 1998), both reasonably in accord with a Norse or immediately post-Norse burial.

Further up the valley is a series of debris cones and fans activated by the 1982 flood (Figs 3.12, 6.5, 6.6). At Lodge Gill [NY 678011: 54.4040, -2.4973]: (Table 3.3, Figs 3.14, 6.11a) is a large debris cone on the eastern valley side, buried by 1982 stony (transitional) debris-flow sediments. This was also a major point of sediment input into Bowderdale Beck during the 1982 flood,

6.11
Middle Bowderdale:
a) The Lodge
Gill tributary to
Bowderdale following
the 1982 flood. Note
the active gully system
and the 1982 debris
cone; b) Leath Gill
and the channel of
Bowderdale Beck
downstream of the
Leath Gill confluence
following the 1982
flood. Note the
braided channel.

resulting in dramatic channel changes at and downstream of the input point (Harvey, 1991). At the input point the channel was diverted westwards, and downstream; a hitherto single-thread channel became wide, shallow and braided. Since then the channel has largely recovered to its pre-flood morphology (Harvey, 2007).

About 400m upstream is another large tributary fan/cone, fed from the east, that of Leath Gill [NY 678006: 54.4000, -2.4958]. This tributary drains a larger catchment than Lodge Gill (Table 3.3) and the 1982 fan exhibited coarse, stony fluvial sediments (Figs 3.14c, 6.11b). Note that the channel of Bowderdale Beck downstream of Leath Gill switched to braided mode following the 1982 storm. Although channel changes resulting from the 1982 flood were less dramatic in Bowderdale than in Langdale (Fig. 6.4), in this part of Bowderdale they were significant, creating new braided reaches. Since then some degree of recovery is evident (see Figs 3.17, 6.4). It is interesting that both Lodge and Leath Gills were major sources of sediment during the 1982 flood, but neither contributed significant sediment during the 2015 flood. The previously unstable reach of Bowderdale Beck that resulted from the 1982 flood was hardly affected by erosion or deposition during the 2015 event. That reach continues to stabilize.

Another 400–600m further up valley are two more tributary fans/cones (Figs 3.12, 6.5, 6.6): Woofler Gill [NY 678003: 54.3964, -2.4963], fed from the west and Wotey Gill [NY 679000: 54.3946, -2.4955], fed from the eastern valley side. Wotey Gill fan, like Leath Gill fan, is a stony fluvially dominant fan, and Woofler Gill fan, fed by a larger catchment, is a wholly fluvial fan (Figs 3.14, 6.6). Woofler Gill contributed no sediment during the 2015 flood. Compare the present stable vegetated fan with the post-1982 situation (see Fig. 3.15b). From Woofler Gill upstream, the channel of Bowderdale Beck is largely a fairly steep bedrock-cut channel with little accommodation space on the valley floor for sediment accumulation. The effects of the 1982 flood were primarily scour.

Continue upstream along the valley floor to the confluence of Randy Gill with Bowderdale Beck [SD 678996: 54.3907, -2.4958]. *You are now in the upper part of the Bowderdale valley.*

Stop 6B.2 Upper Bowderdale [SD 678996: 54.3907, -2.4958]
There is no depositional feature at the confluence of Randy Gill, from the east, with Bowderdale Beck. The catchment of Randy Gill is relatively large, with little evidence of 1982 hillslope erosion, so presumably the yield was primarily of floodwater, preventing deposition at the tributary junction.

Within this reach of the main valley are two further 1982 fans/cones, both involving dominantly fluvial fan deposition (Table 3.3): at Hazel Gill (Fig. 3.14; [SD 678996: 54.3907, -2.4957]) and at Rams Gill [SD 678988: 54.3840, -2.4965] (Fig. 6.6). Channel sedimentation was restricted to the immediate vicinities of the confluences. Further up the valley are two more tributary gully systems: Low Green Gill [SD 679985: 54.3799, -2.4941] and High Green Gill

[SD 682982: 54.3775, -2.4910], both on the eastern valley side. Both catchments had some gully erosion during the 1982 storm, but yielded insufficient sediment to influence the channel of Upper Bowderdale. High Green Gill is actually the main headwater of Bowderdale Beck. Bowderdale ends in the col above Cautley 200m upstream, marking the original course of Bowderdale Beck pre-glacial(?) headwaters, prior to (late Pleistocene) glacial capture by Cautley Holme Beck (see Chapters 2 and 7A).

The current head of Bowderdale contrasts markedly with the three heads of the Langdale system, the open col contrasting with the well-developed valley-head dendritic gully systems of East, Middle and West Grains at the head of Langdale (see previous section, 6A).

This marks the end of the Bowderdale field trip – return back down the valley to your vehicle.

C Weasdale

I do not give a full account of the geomorphology of Weasdale, the next valley east of Bowderdale. It was much less affected by the 1982 storm than either Langdale or Bowderdale, and there has been much less research work on the Holocene sequence there than elsewhere in the Howgills. However, for a feel of the general geomorphology of the northern Howgills, Weasdale may be worth half a day's visit. The total hiking distance of 7km (round-trip) is markedly less than for either Langdale or Bowderdale. The route is shown on the Google Earth image (Fig. 6.1).

Access: At Tranna Hill (Newbiggin-on-Lune) [NY703054: 54.4422, -2.4601], *on the A685 about 10km east of Tebay (or c.1km west of Ravenstonedale), turn south onto the minor road signed Weasdale. Drive south and southwest for c.2km to the hamlet/farm of Weasdale, just beyond Weasdale Nurseries. Park here* [NY 691038: 54.4286, -2.4788]. *Walk up the path south from Weasdale farm on the eastern side of the Weasdale valley. After c.500m cross an incised tributary stream (Pinksey), then head down towards the west and onto the valley edge. At this point you pass from improved meadows into unfenced rough grazing land. It is probably best to keep above the valley floor for another 250m or so, then to descend onto the valley floor. You are now at the junction of Little Swindale and Weasdale* [NY 688026: 54.4180, -2.4820]: *(Fig. 6.1).*

Stop 6C.1 Lower Weasdale [NY 688026: 54.4180, -2.4820]

The Little Swindale tributary drains a nondescript catchment from the east, with little evidence of historic gullying, only minor drainage lines on the hillslopes. The channel of Little Swindale Beck is incised below the lower hillslopes. Small, now-stable gully scars cut into the slopes bounding the incision. At the tributary junction of Little Swindale Beck with Weasdale Beck there is what may be an ill-defined low-angle fan, now dissected by the tributary stream. Between here and the junction with Great Swindale [NY 688018: 54.4111, -2.4820] Weasdale Beck is a wandering single-thread channel set below smooth solifluction surfaces. On the west valley side there are ghost drainage lines across this surface, but no evidence of significant historic, let alone modern, erosion.

The Great Swindale tributary is different, more like the tributary valleys in the other northern Howgill catchments. On the valley sides linear gullies cut into the solifluction surfaces, below which the stream is incised. Small gullies/scars pock the slower slopes. There is little sign of much modern active erosion, so it must be assumed that the larger features, at least, are equivalent to the tenth-century features identified elsewhere. At the head of the Great Swindale valley are more complex (now mostly stable) gully networks. In its headwaters the floor of Great Swindale valley opens out and is formed of fluvial sediments, presumably dating from when the gullies were actively supplying sediment from the hillslopes. A stable vegetated fan occurs at the Great Swindale/Weasdale Beck confluence.

Continue upstream along the Weasdale valley into upper Weasdale **[NY 690008: 54.4018, -2.4787]**.

Stop 6C.2 Upper Weasdale [NY 690008: 54.4018, -2.4787]

The main valley is similar here to that downstream from the Great Swindale tributary junction, albeit narrower with a smaller channel and set less deeply below the valley sides. The linear features dissecting the western solifluction slope are a little deeper (Fig. 6.12a) and a few show signs of modern erosion.

Upper Weasdale is an arcuate valley head, similar to West, Middle and East Grains in Langdale (see above, Section 6A). It is heavily gullied by dendritic

6.12a Looking down Weasdale from Weasdale Head;

6.12b Looking west across the gully systems at Weasdale Head.

gully networks (Fig. 6.12b), presumably of tenth-century origin. On some of the gullied slopes some modern erosion is apparent. There is a fan-like depositional feature where the three headwater gully networks converge.

This marks the end of the Weasdale field hike: return back down the valley to your vehicle.

CHAPTER 7

The Southern Howgills

Introduction

In this chapter I deal with several sites within the southern Howgills. Because our research work has been primarily focused on Carlingill (Chapter 5) and the northern Howgills (Chapter 6), the treatment here will be less detailed and more general than in the previous two chapters. Within this chapter I focus on three areas (Fig. 7.1), each with its own distinctive characteristics. These areas are: (A) Cautley, with the evidence of local glaciation (see also Chapter 1); (B) the small, steep catchments in the Sedbergh area in the far south of the Howgills, and (C) Chapel Beck in the south-western Howgills, where we have only recently begun to piece together the Holocene sequence (Chiverrell *et al.*, 2008).

Highlights: Glacial erosion and drainage diversion at Cautley; Small, steep incised valleys near Sedbergh; a large valley system, but with only limited erosion at Chapel Beck.

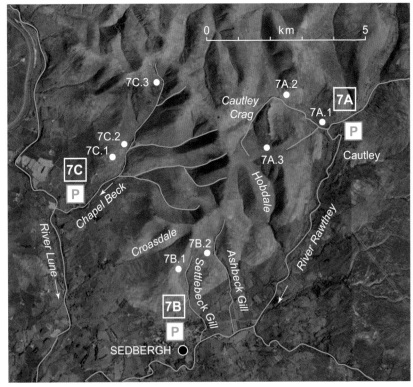

7.1 Google Earth image of the southern Howgills, showing the locations of parking areas for the field sites 7A: for sites 7A.1–3; 7B: for sites 7B.1 and 7B.2; 7C: for sites 7C.1–3.

A Cautley (Figs 7.2, 7.3; see also Figs 1.6, 4.7)

Introduction: This excursion could either be a simple, fairly gentle walk of less than 3km round trip (**Stops 1** and **2** only), or a longer excursion involving clambering up very steep rugged terrain ultimately to an elevation of over 650m (2150ft). You would need to be fit and properly equipped!

Directions: On the A683 Sedbergh to Kirkby Stephen road, about 7km NE of Sedbergh is the Cautley Cross Keys Temperance Inn [SD 697969: 54.3670, -2.4660]. *Just beyond the inn is a barn, beyond which is roadside parking on the left. Park here, then walk behind the barn to a footbridge across the River Rawthey. Cross the footbridge and turn left along a path above the river bank to a point above where Cautley Holme Beck joins the River Rawthey from the west* [SD 693968: 54.3656, -2.4730]. *This is Stop 7A.1.*

Stop 7A.1 Confluence of Cautley Holme Beck with the River Rawthey [SD 693968: 54.3656, -2.4730]

This site gives an overview of the Cautley area. Look to your northwest and you will see the panorama of the Cautley cirque and surrounding area (see Fig. 4.7). Cautley cirque supported a local glacier after the late Pleistocene ice sheet had melted, though whether this was at the close of the main glaciation (20–18ka BP) or during the so-called Loch Lomond re-advance (11–10ka BP) when there were cirque glaciers in the Lake district (Sissons, 1980), is uncertain (see Chapter 1). The local extent of the glacier, however, is fairly clear. The gentle slope above you to your right is clearly a solifluction slope probably formed contemporaneously with the local Cautley glacier, but it is outside the local glacial limit. In front of you in a WNW direction, the

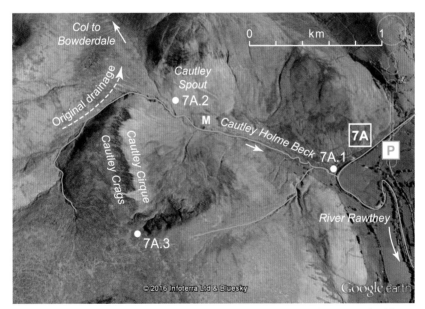

7.2 Google Earth image of the Cautley area, showing stop locations; see also Figure 4.7 for general photo of the Cautley area (M is the mound – see text for possible explanations).

mound above the valley floor is possibly a terminal moraine of the local glacier (Fig. 4.7). However, it has been interpreted as a glacially overrun slope failure originally derived from the slope above (Mitchell, 1991; Jarman *et al.*, 2013). Above this mound are scree slopes on the flanks of Yarlside Hill, outside the limits of the Cautley glacier.

To the left (west) of the flanks of Yarlside, partly hidden by Yarlside itself, is the col into the current head of Bowderdale (see Chapters 1, 6). Originally the Bowderdale headwaters drained the area behind (west of) Cautley cirque (Fig. 4.7), but were diverted to the present course into Cautley Holme Beck by glacial erosion of the cirque itself (see Chapter 1). The resulting scenery is spectacular. Red Gill Beck (now the main headwater stream of Cautley Holme Beck) cascades into the Cautley valley in a series of waterfalls, Cautley Spout (Fig. 7.3a), supposedly the highest waterfall in England. To the left (S) of Cautley Spout the cliffs and screes of the backwall of the cirque are visible (see below, Stop 7A.2), in front of which, outside the limits of the cirque glacier, are the flanks of Coonthard Brow, above the Rawthey valley.

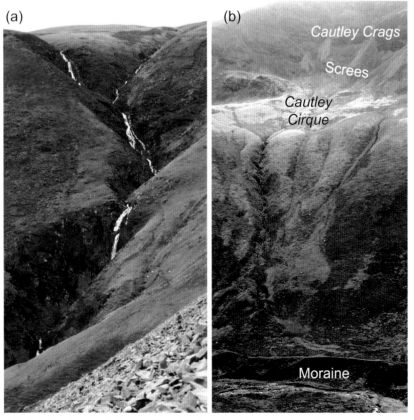

7.3 Cautley cirque: **a)** Cautley Spout waterfalls; **b)** Looking into Cautley cirque from the foot of Cautley Spout. (See also Fig. 1.6 for view north from above Cautley cirque.)

From this location (*Stop 7A.1*) other features of the geomorphology are evident. The eastern flanks of Yarlside (to your northeast), as well as those of Coonthard Brow (to your southwest) are clearly outside the limits of the local cirque glacier and are blanketed by solifluction deposits. Gully systems (presumably as elsewhere in the Howgills, related to tenth-century vegetation changes), dissect these slopes. The River Rawthey, in front of you (to your SE), has here a single-thread alluvial channel, but beyond the large bend in front of you it is incised into bedrock. In this part of the valley the river is flanked by terraces, however we do not know the details of the sequence here.

Continue along the track, now heading NW into the valley of Cautley Holme Beck, on the north side of and above the stream. Continue behind the mound to a point above the small tributary stream that descends from the Bowderdale col, just below the lowest part of Cautley Spout waterfalls. This is **Stop 7A.2** [SD 683975: 54.3719, -2.4888]*, giving a view towards Cautley cirque.*

Stop 7A.2 Viewpoint into Cautley cirque [SD 683975: 54.3719, -2.4888]

From this viewpoint the overall morphology of the cirque can be appreciated (Fig. 7.3b). To the right (west) is the backwall of the cirque, cliffs above scree tongues. The floor of the cirque basin is irregular, but culminates in front of you in a marked steepening over the lip of the cirque, which during glaciation would have formed an ice fall. At the foot of the steep slope is a morainic ridge, presumably that of a recessional stage (Fig. 7.3b). Nearer at hand and to your right, you have a close view of Cautley Spout waterfall.

This forms the end of the basic excursion in the Cautley area. Unless you are taking the latter part of the excursion, return to your car the way you came. However, if you are taking the latter part of the excursion, make sure you are fit. Make sure you are adequately clothed and shod. Do not attempt this part of the excursion in rainy or cloudy weather.

If taking this part of the excursion, follow the very rough path ('path' is a bit of a generous term here!), that keeps to the north (right) of Red Gill Beck. Near the top of the worst of the climb cross a small tributary [at SD 679975: 54.3724, -2.4948]*, turn SW, then cross the incised channel of Red Gill Beck and continue to climb up the ridge. Keep away from the hazardous cliff tops to your left (east). Continue up the ridge to the summit of Great Dummocks Hill* [SD 681963: 54. 3617, -2.4929]*. This is* **Stop 7A.3**, *above the backwall of Cautley cirque.*

Stop 7A.3 Above the backwall of Cautley cirque [SD 681963: 54. 3617, -2.4929]

In addition to a spectacular 360-degree panoramic view north over the central Howgills, west towards the flanks of the Lake District, east and south into the Yorkshire Dales National Park, you have a superb local view of Cautley Cirque and associated geomorphology (see Fig. 1.6). Looking into the cirque, the cliffs below you and on the left (western) flank are footed by scree tongues. The irregular floor of the cirque is below you. Beyond the cirque is the col into the head of Bowderdale (see Chapters 1, 6), through which the original Cautley drainage flowed into Bowderdale, prior to its glacial diversion into its present course into the Rawthey.

*This is the formal end of the Cautley part of the excursion. The best way to return is to head east then ENE down the steep hillslope (keeping well away from the cliff edge on the southern margin of the cirque), directly down the slope to **Stop 7A.1**, the confluence zone of Cautley Holme Beck and the River Rawthey. Then go back to your car the way you came along the river. From here drive SW along the A683 for about 7km to Sedbergh. Park within Sedbergh.*

B The Sedbergh area – Settlebeck and Ashbeck Gills

The SE flank of the Howgills is a relatively steep slope, parallel with the Rawthey valley. It is probably coincident with a WSW–ENE orientated fault that separates the Rawthey valley terrain on the basal Carboniferous conglomerate, from the Silurian rocks of the Howgills themselves (see Fig. 1.4).

Midway between Cautley and Sedbergh, the channel of the River Rawthey is cut into bedrock, exposing the basal Carboniferous conglomerate. There are three drainage basins within the Howgills in this area. Hobdale Beck (Fig. 7.1) is trenched into bedrock upstream of its confluence with the Rawthey, but the upper basin is a typical Howgill drainage basin, whose slopes are dissected by a set of now stabilized tenth-century(?) gullies. The two other basins, Ashbeck and Settlebeck (Fig. 7.4), nearer to Sedbergh, are smaller. In both,

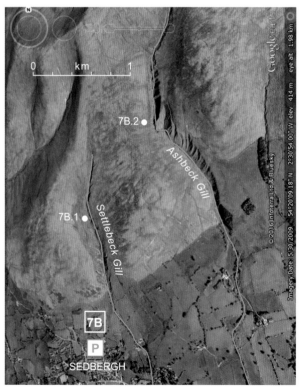

7.4 Google Earth image of the Sedbergh area and the southern Howgills, showing locations of Settlebeck and Ashbeck Gills and Stops 7B.1 and 7B.2.

the main channels are cut into solifluted glacial deposits, which in each case form benches within the valleys. Ashbeck Gill, though, cuts through the drift deposits into the underlying bedrock. In both valleys small gullies, both active and now stabilized dissect the slopes where the channels have been incised into the solifluction benches. Settlebeck Gill (*Stop 7B.1*) is easily accessible from Sedbergh. A short visit (less than two hours), involving a short hike (of less than 3km, round trip) would allow you to appreciate the geomorphology.

Directions: Park within Sedbergh, then head on foot to the NE corner of the town [SD 658924: 54.3260, -2.5236]. *From here take the lane that heads NE towards Settlebeck Gill. Follow the stream into the Settlebeck valley within the Howgills. Once within the valley itself, clamber up to a bench-like form on the west side of the valley – this is Stop 7B.1* [SD 659932: 54.3313, -2.5253].

Stop 7B.1 Settlebeck Gill [SD 659932: 54.3313, -2.5253] (Figs 7.4, 7.5a)
From here there is a view of the whole valley system: the solifluction benches, the modern channel incised into them, the active scars and gullies and the older stabilized gullies cut into the incisional slopes, and down-valley the former valley floor preserved as a terrace above the modern channel (see Fig. 7.5a).

Either: Retrace your steps back to Sedbergh or alternatively, if you wish also to see Ashbeck Gill, extending the walk to a total of c.5km round trip including a moderate climb of c.110m (c.350ft), continue walking up Settlebeck Gill valley to the head of the solifluction terrace [SD 660936: 54.3368, -2.5242]. *Then head ENE over the shoulder to the north of Crook Hill* [SD 665937: 54.3395, -2.5240]. *From here descend towards the NNE into the Ashbeck valley to a point above the deeply incised stream. This is Stop 7B.2* [SD 666938: 54.3394, -2.5149].

Stop 7B.2 Ashbeck Gill [SD 666938: 54.3394, -2.5149]
The topography of the Ashbeck Gill valley (Fig. 7.5b) is similar to that of the Settlebeck Gill valley, but in some ways better developed. The slopes are mantled by solifluction deposits, especially lower down the valley. The channel is deeply cut into these sediments, locally producing exposures of solifluted till over undisturbed till. The valley walls are marked by small, steep, now mostly stabilized gully systems. From mid-valley upstream the modern channel cuts through the solifluction deposits and is incised into bedrock below (see Fig. 7.5b).

End of this part of the excursion – return to Sedbergh. Descend down the valley, keeping to the west side of Ashbeck Gill, to a set of buildings at [SD 670930: 54.3319, -2.5093] *(labelled Gill on OS maps). From there, there is a footpath along the southern flank of the Howgills back to Sedbergh.*

C Chapel Beck – the southwestern Howgills

The geomorphology of Chapel Beck (Figs 7.6, 7.7a,b) has only recently come under intensive study, under the leadership of Richard Chiverrell of the University of Liverpool (Chiverrell *et al.*, 2008). Chapel Beck is the larger of

7.5 The southern Howgill valleys: a) Settlebeck Gill: note how the valley cuts through the solifluction surface and deposits into the underlying glacial till. Note also the incisional terrace. b) Ashbeck Gill: note how the head of this channel cuts into bedrock.

the two main SW Howgill basins (Croasdale and Chapel Becks). In contrast with Carlingill (see Chapter 5), in neither of these basins is there much active modern erosion, but up to a point the solifluction hillslopes are dissected by mostly now stable, presumably tenth-century (?) gully systems. In Chapel Beck there are small fans and cones at the tributary junctions with the main streams, from which ^{14}C dating has tightened our understanding of the timing of hillslope erosion in the Howgills (Chiverrell et al., 2008).

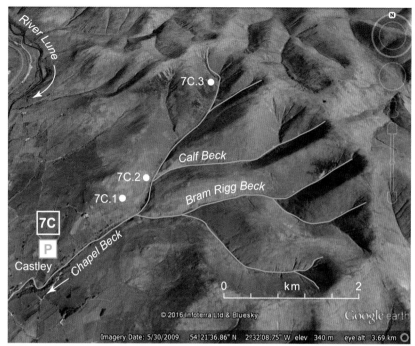

7.6 Google Earth image of the Chapel Beck area, showing locations of field sites 7C.1–3.

Directions: From Sedbergh take Howgill Lane out of the NW corner of the town north towards Borrowbridge and Tebay. Continue on this road beyond Chapel Bridge [SD 635950: 54.3496, -2.5643] *for another kilometre to the road junction with Beckfoot Lane (from the west)* [SD 633958: 54.3570, -2.5672]. *If coming from Tebay and the north, take Howgill Lane south from Low Borrowbridge* [NY 607014: 54.4068, -2.6067] *for about 6km to the junction of Howgill and Beckfoot Lanes* [SD 633958: 54.3496, -2.5643]. *At the junction of Howgill Lane and Beckfoot Lane (from the west), turn east here onto a minor lane to the second set of farm buildings: Castley* [SD 637958: 54.3564, -2.5584]. *Park outside these buildings [From here you should probably allow up to half a day to visit Stops* **7C.1**, **7C.2** *and* **7C.3** *within the upper part of the Chapel Beck valley, more if you wish to explore the other tributary valleys]. From Castley walk further east along the track (the lane you have driven along degenerates into a track) into the Chapel Beck valley, to a point opposite a large streamside scar that exposes stratified solifluction deposits (see Fig.1.8). This is* **Stop 7C.1** *(Figs 7.6, 7.7a,b)* [SD 644960: 54.3582, -2.5503].

Stop 7C.1 Chapel Beck, overview [SD 644960: 54.3582, -2.5503]
From here you have an overview of the Chapel Beck system (Fig. 7.7a), cut below the Howgill summit erosion surface (See Chapter 1; Fig. 1.9). The Chapel Beck system comprises a multi-branched stream network. The landform assemblage is broadly similar to that in other Howgill valleys (see Google Earth image, Fig 7.6). The stream channels are cut into the base of solifluction-mantled hillslopes. Note in front of you the superb section in bedded solifluction sediments overlying glacial till (see Fig. 1.8a,b).

(a)

(b)

7.7 Chapel Beck: a) Overview of the Chapel Beck catchment. Note the exposure of stratified Late Pleistocene solifluction sediments on the margins of Chapel Beck (for more detail see Fig. 1.8a,b). Note also the relatively shallow slope stripping rather than deep hillslope gullies within this catchment. b) A shallow gully and debris cone within the Chapel Beck valley.

The arcuate valley heads in the Chapel Beck system are somewhat less intensively gullied by the ?tenth-century (now-stable) gully networks than those of other Howgill valleys, (particularly Carlingill and the northern valleys (compare Fig. 1.9 with Fig. 2.5; see also Chapters 5 and 6). There is relatively little active modern erosion, much less for example than in the Carlingill drainage. There is only one large gully-foot tributary-junction fan, near the head of Chapel Beck at Stranger Gill. Otherwise the tributary junctions are marked by much smaller cones. This might relate to limited sediment supply or perhaps to limited accommodation space on the relatively narrow valley floor. Sections in five of these small cones expose buried soils below the cone sediments. These soils have yielded radiocarbon dates that enable the age of overlying cone sediments to be estimated (Chiverrell *et al.*, 2008). One is a very recent date that tells us little. The other four show a wide range of ages. A buried soil within the multi-surfaced cone at the mouth of Calf Beck yields a date of 2035 +/-35 BP. Another buried soil from the simple cone at the confluence of Swarth Graves and Bram Rigg Becks (in the south of the basin [SD 650957: 54.3563, -2.5491]) yields a 'Norse' date of 1090 +/-35 BP, and two small cones at the head of Chapel Beck itself, Long Rigg Gill and White Fell Gill, yield younger dates of 280 +/-35 BP and 275 +/-35 BP respectively.

*From the Chapel Beck overlook (**Stop 7C.1**) the track descends into the valley, reaching the valley floor more or less where Calf Beck joins the main stream. This is **Stop 7C.2** [SD 645964: 54.3614, -2.55459].*

Stop 7C.2 Chapel Beck/Calf Beck confluence [SD 645964: 54.3614, -2.5459]

The Calf Beck confluence is marked by a multi-surfaced cone, yielding a radiocarbon date that indicates burial sometime after 2035 BP.

*From Calf Beck head upstream alongside the main stem of Chapel Beck to the head of the valley to **Stop 7C.3** [at SD 655979: 54.3760, -2.5333].*

En route, note the cones issuing from Long Rigg Gill, on the left at [SD 649969: 54.3664, -2.5418] and White Fell Gill, on the right at [SD 652973: 54.3720, -2.5337], both of which yielded relatively young dates (see above). Also note the larger (undated) fan at Stranger Gill (on the right at [SD 654975: 54.3686, -2.5360]). Note en route the very limited extent of modern erosion within the Chapel Beck system, although there is the occasional active gully and debris cone (Fig. 7.7b).

Stop 7C.3 Head of Chapel Beck [SD 655979: 54.3760, -2.5333]

From this point there is a view into the head of the Chapel Beck valley. Note the limited Norse (?) age gullying. Note also the limited extent of modern erosion: what there is occurs primarily on the side-slopes of the older gullies. Note also, that I have not visited this area since the 2015 storm event, so there may be effects of that storm of which I am unaware!

End of itinerary – return to Castley the way you came.

CHAPTER 8

Final reflections

In this final chapter we need to consider several questions: (1) How typical or representative are the Howgills of other upland areas in Britain? (2) How 'special' are the Howgills? Are there aspects of geomorphology that are better seen here than anywhere else? And (3) What are the remaining questions and uncertainties related to the specific aspects of the geomorphology of the Howgills?

(1) In answer to the first question, how typical the Howgills are of upland Britain as a whole, the key lies in the glacial sequence. During the last Pleistocene glaciation the Howgills were affected by the regional ice sheet and did not function as a glacial source area. In this way they are similar to parts (but probably not to the central parts) of the Southern Uplands of Scotland, and also to some of the flanks of the Pennines (e.g., the Bowland Fells of the Lancashire/Yorkshire border). The main mass of the Pennines is rather different. There, lithological/structural controls of the landscape are dominant. There may be some similarities between the Howgills and parts of central Wales, but the local relief there tends to be less. The Howgills are not representative of glacial source areas such as the Scottish Highlands, the Lake District or North Wales.

Glacial drainage diversions are characteristic of all glaciated areas, therefore are not distinct to the Howgills or to the similar areas identified above. However, the examples here, both large (the Lune at Tebay?) or small (Bowderdale headwaters and Blakethwaite) are exceptionally clear. The Pleistocene glacial sequence conditioned the landscape for the geomorphic activity that followed. The lengthy period of permafrost during the late Pleistocene rendered the Howgill hillslopes prone to solifluction, a characteristic of the areas outside the Late Pleistocene Loch Lomond limits (see Chapter 1; see also Harrison, 1996). Solifluction conditioned the landscape for later geomorphic activity. In the similar areas identified above, there are common aspects to the Holocene sequence. Early incision into the base of solifluction slopes or into late Pleistocene valley floors created the highest of the post-glacial stream terraces, as in many other areas in upland Britain.

For the Holocene sequence itself, variations in sediment supply, whether climatically driven or, as is increasingly the case later in the Holocene, influenced by human activity (Coulthard and Macklin, 2001), create both similarities and differences between the various parts of upland Britain. There is some similarity between the Howgills and, for example, the Southern Uplands, the Bowland Fells (Chiverrell *et al.*, 2007), and perhaps central Wales. However, Holocene hillslope gullying and associated alluvial fan deposition are also characteristic of most former glaciated areas in upland Britain, including the former glacial source areas such as the Scottish Highlands and the Lake District (Rumsby and Macklin, 1994; Tipping and Halliday, 1994; Macklin *et al.*, 1992; Brazier *et al.*, 1988; Tipping *et al.*, 1999; Macklin and Lewin, 2003; Lewin *et al.*, 2005; Chiverrell *et al.*, 2007). Late Holocene hillslope gullying and alluvial fan deposition are exceptionally well developed in the Howgills.

Modern processes in the Howgills are NOT typical of other areas in upland Britain, at least in their extent and intensity. Few, if any, other areas show the same intensity of modern processes: active hillslope gullying, and the coupling relationships (demonstrated in Chapter 3) between slope (gullying) and channel processes.

(2) To answer the second question posed at the beginning of this chapter, i.e., the distinctiveness/ special character of the Howgill geomorphology, in some ways the answer is the reverse of the aspects described above. Two characteristics stand out. Firstly, historical (primarily tenth-century) hillslope gullying, involving the coupling relationships between gullying, fan and channel processes, are exceptionally well developed here. Similarly, Holocene alluvial fans are particularly well developed in the Howgills. Secondly, modern active gullying, supplying sediment to coupling zones and the modern channel system is exceptionally clear here. As a result, the channel response to spatial and temporal variations in sediment supply is also clear.

Added to the characteristics described above, the openness of the Howgill landscape is important, combined with the steepness of the valley side-slopes. These properties potentially intensify the geomorphic processes, and potentially create spatial variability in geomorphic threshold exceedance. Furthermore, there is almost no major *direct* human modification of the landscape. There is no human-related built infrastructure. There are no settlements, no roads – there are not even walls within the main part of the Howgill Fells. For the late Holocene sequences, influenced by climate–vegetation/land use interactions, and in the context of active modern processes, there is little *direct* human effect on the geomorphic system. Within the British context, apart perhaps from the more remote parts of the Scottish Highlands, the Howgill Fells are as close as one can get to a 'natural' landscape. This characteristic makes the Howgill Fells an ideal field laboratory for the study of the geomorphology of slope-stream sediment systems.

(3) The third question raised in the introduction to this chapter relates to the questions and uncertainties that remain in relation to the geomorphology of the Howgill landscape. The first set of questions/unknowns relates to the pre-glacial landscape. We know virtually nothing of the erosion-surface story apart from their general slope in the northern Howgills towards the north and the Eden drainage. We know nothing of the timing, chronology, nor of the origins of the erosion surfaces (peneplains, pediplains, etchplains? – see Chapter 1). It is difficult to envisage from where answers to these questions might come – probably not from the Howgills at all! A related question involves the drainage evolution, and the eventual switch from northerly drainage towards the Eden to southward drainage towards the lower Lune (see Chapter 1). Was it a simple headwards capture process or was there an intermediate stage whereby the Lowther headwaters captured the northern Howgill streams through the Shap gap before eventual glacial diversion southwards? The answers to these questions may relate to the effects of earlier (mid-Pleistocene) glaciations.

Pleistocene glaciation is not my research field, but it seems to me that there are unanswered questions relating even to the last glaciation. How does the Lake District/Northern England glacial sequence relate to the standard Scottish chronology? Even more obscure are the possible effects of earlier glaciations. Are there pockets of glacigenic sediments within the region as a whole that could be related to such glaciations? If so, are there any inferences that can be made relating to the effects of such glaciations on the landscape?

A second set, of perhaps less intangible questions, relates to the closing stages of the last glaciation. What were the spatial patterns of glacial erosion and deposition? What was the role of meltwater? What was the timing of deglaciation, and particularly what was the duration of periglacial conditions? What does then become fairly clear is the late Pleistocene sequence, involving the creation of the hillslope solifluction surfaces and at the same time the deposition of gravels on the then valley floors to form what are now the high terraces of the Howgill streams. One question here, which we assume is straightforward, but have not studied in any detail, is the relation between the high terraces within the Howgill valleys and the terrace sequences of the main Lune and Rawthey valleys.

Within the Howgills the Holocene sequence is becoming clearer, as is its relation to that in neighbouring areas, thanks to the recent work led by Richard Chiverrell (Chiverrell *et al.*, 2007, 2008). That work is also addressing the question of climatic vs. vegetation/land-use influence. It is now clear that rather than only one major tenth-century phase of gullying and fan deposition, several other minor phases can be identified. One local question in relation to this theme would be the spatial pattern of responses. Can topographic thresholds be identified for these other slope-erosional events? If so, how do they differ from those related to the major tenth-century phase? Do they show differences among the Howgill valleys?

In relation to modern processes, two questions appear to be significant. First is the relation between hillslope destabilization and vegetative healing. This has importance in relation to the coupling characteristics, and therefore has implications for stream channel geometry and processes. It is also related to a second question, that concerning the significance of major flood events and their role in relation to 'normal' processes. It is quite obvious that the effects of the two extreme events that have occurred over the last 40 years were markedly different – 1982 dominated by intense slope/gully erosion and tributary fan/cone deposition, with some effects on the main channels – 2015 dominated by stream-flood processes especially influencing the main channels. What is the significance of contemporary postulated climatic change?

Whatever the answers to the geomorphic questions posed above, the Howgills remain a superb field site in which to address some of these research problems. They also form a beautiful and fascinating landscape (Wainwright, 1972), little touched by modern development – an important landscape to be conserved!

APPENDIX 1

Published map coverage

Published map cover is variable. Of the two regular scales of topographic map cover published by the Ordnance Survey, the Landranger Series at the 1:50,000 scale is not very convenient. Three sheets are needed to cover the Howgill Fells: Sheets 91 Appleby, 97 Kendal and Morecombe, and 98 Wensleydale and Upper Wharfedale. However, at the much more useful, Explorer Series 1:25,000 scale the Howgills are covered on one sheet: Sheet OL19 Howgill Fells and Upper Eden Valley.

For geology maps, published by The British Geological Survey, the situation is good, with most of the Howgills, except the far northeast of the Howgills, published on one 1:50,000 Solid and Drift Geology sheet (Sheet 39: Kendal). For the far northeast of the Howgills you would need Sheet 40, Solid and Drift Geology, Kirkby Stephen.

APPENDIX 2

Particle size: the phi scale

Particle size is shown in phi φ units, a negative logarithmic scale, the conventional scale used for sediment sizes,

$$\text{where } \varphi = - \log_2 d$$

(d is the particle diameter in mm). See below for conversions between mm and phi units.

Mm	0.01	0.063	0.125	0.25	0.5	1	2	4
Phi units	6.64	4	3	2	1	0	-1	-2

Sand sizes range from -1 to 4 phi φ, silt from 4 to 9 phi φ, clay from 9 to 12 phi φ.

APPENDIX 3

Grid References and GPS co-ordinates

National grid references and GPS co-ordinates for main sites related to the field excursions (**Chapters 4–7**)

Chapter 4 The Reconnaissance Trip
4.1 Dillicar Common:	SD 613988	54.3861, -2.5955
4.2 Lune Gorge overlook	NY 607007	54.4005, -2.6053
4.3 Lune's Bridge, Tebay	NY 613028	54.4197, -2.5979
4.4 Orton Scar	NY 628098	54.4827, -2.5764
4.5 Sunbiggin Tarn	NY 675078	54.4650, -2.5021
4.6 The Dent Fault	SD 730996	54.3897, -2.4190
4.7 Cautley	SD 697969	54.3670, -2.4660

Chapter 5 The Western Howgills
Howgill Lane – parking spot	SD 624995	54.3900, -2.5800
5.1 Grains Gill viewpoint	SD 627997	54.3913, -2.5751
5.2 Grains Gill gullies	NY 628003	54.3953, -2.5741
5.3 Along Carlingill valley	SD 627997	54.3915, -2.5735
	to SD 635944	54.3895, -2.5678
5.4 Middle Carlingill	SD 635994	54.3888, -2.5629
	to SD 640993	54.3871, -2.5568
5.5 Above Carlingill Gorge	SD 644994	54.3881, -2.5512
5.6 Blakethwaite	SD 648998	54.3929, -2.5440

Chapter 6 The Northern Howgills
Longdale – parking spot	NY 645051	54.4397, -2.5493
6A.1 Burnt Gill, Langdale	NY 662016	54.4081, -2.5212
6A.2 Thrush Gill, Langdale	NY 666012	54.4044, -2.5149
6A.3 Upper Langdale	SD 666997	54.3920, -2.5154
Bowderdale – parking spot	NY 683043	54.4322, -2.4922
6B.1 Greencomb Gills	NY 675017	54.4107, -2.5008
6B.2 Upper Bowderdale	SD 678996	54.3907, -2.4958
Weasdale – parking spot	NY 694038	54.4286, -2.4788
6C.1 Lower Weasdale	NY 688026	54.4180, -2.4820
6C.1 Upper Weasdale	NY 690008	54.4018, -2.4787

Chapter 7 The Southern Howgills
Cautley – parking spot	SD 697969	54.3671, -2.4659
7A.1 Below Cautley	SD 693968	54.3661, -2.4727
7A.2 Nr Cautley Spout	SD 683975	54.3719, -2.4888
7A.3 Above Cautley Crags	SD 681963	54.3617, -2.4929
Sedbergh	SD 658924	54.3260, -2.5236
7B.1 Settlebeck Gill	SD 659932	54.3313, -2.5253
7B.2 Ashbeck Gill	SD 666938	54.3394, -2.5149
Chapel Beck – parking spot	SD 637958	54.3564, -2.5584
7C.1 Chapel Beck overview	SD 644960	54.3582, -2.5503
7C.2 Chapel Beck/Calf Beck	SD 645964	54.3614, -2.5459
7C.3 Head of Chapel Beck	SD 655979	54.3760, -2.5333

REFERENCES

Al Farraj, A. and Harvey, A.M. (2010) Influence of hillslope-to-channel and tributary-junction coupling on channel morphology and sediments: Bowderdale Beck, Howgill Fells, NW England. *Zeitschrift fur Geomorphologie* **54**, 203–224.

Brazier, V., Whittington, G. and Ballantyne, C.K. (1988) Holocene debris-cone evolution in Glen Etive, Western Grampian Highlands, Scotland. *Earth Surface Processes and Landforms* **13**, 525–531.

Chiverrell, R.C. (2001) A proxy record of Late Holocene climate change from May Moss, northeast England. *Journal of Quaternary Science* **16**, 9–29.

Chiverrell, R.C. and Thomas, G.S.P. (2010) Extent and timing of the Last Glacial Maximum (LGM) in Britain and Ireland: a review. *Journal of Quaternary Science* **25**, 535–549.

Chiverrell, R.C., Harvey, A.M. and Foster, G.C. (2007) Hillslope gullying in the Solway Firth–Morecambe Bay region, Britain: responses to human impact and/or climatic deterioration? *Geomorphology* **84**, 317–343.

Chiverrell, R.C., Burke, M.J. and Thomas, G.S.P. (2016) Morphological and sedimentary responses to ice mass interaction during the last deglaciation. *Journal of Quaternary Science* **31**, 265–280.

Chiverrell, R.C., Harvey, A.M., Hunter (née Miller), S.Y., Millington, J. and Richardson, N.J. (2008) Late Holocene environmental change in the Howgill Fells, Northwest England. *Geomorphology* **100**, 41–69.

Coulthard, T.J. and Macklin, M.G. (2001) How sensitive are river systems to climate and land-use changes? A model-based evaluation. *Journal of Quaternary Science* **16**, 347–351.

Cundill, P.R. (1976) Late Flandrian vegetation and soils in Carlingill valley, Howgill Fells. *Transactions of the Institute of British Geographers* **4**, 301–309.

Cundill, P.R. (2000) Deforestation and erosion episodes in the Howgill Fells. In Hodgkinson, D., Huckerby, E., Middleton, R. and Wells, C.E. (eds) (2000) *The lowland wetlands of Cumbria*. Lancaster: Lancaster Archaeological Unit, 323–326.

Dunsford, H.M. (1998) *The response of alluvial fans and debris cones to changes in sediment supply: upland Britain*. PhD Thesis, University of Durham, 382pp.

Edwards, W. and Trotter, F.M. (1954) *The Pennines and Adjacent Areas*, British Regional Geology. London: HMSO, 86pp.

Evans, D.J.A., Livingstone, S.J., Vieli, A. and Cofaigh, C.O. (2009) The palaeoglaciology of the central sector of the British and Irish Ice Sheet: reconciling glacial geomorphology and preliminary ice sheet modelling. *Quaternary Science Reviews* **28**, 739–757.

Harrison, S. (1996) Paraglacial or Periglacial? The sedimentology of slope deposits in Upland Northumberland. In Brooks, S.M. and Anderson, M.G. (eds) (1996) *Advances in Hillslope Processes*, Volume 2. Chichester: Wiley, 1197–1218.

Harvey, A.M. (1974) Gully erosion and sediment yield in the Howgill Fells, Westmorland. In Gregory, K.J. and Walling, D.E. (eds) (1974) *Fluvial Processes in Instrumented Watersheds*, Institute of British Geographers, Special Publication **6**, 45–58.

Harvey, A.M. (1977) Event frequency and channel change. In Gregory, K.J. (ed.) (1977) *River Channel Changes*, Chichester: Wiley, 301–315.

Harvey, A.M. (1986) Geomorphic effects of a 100-year storm in the Howgill Fells, Northwest England. *Zeitschrift fur Geomorfologie NF* **30**, 71–91.

Harvey, A.M. (1987a) Seasonality of processes on eroding gullies: a twelve-year record of erosion rates. In Goddard, A. and Rapp, A. (eds) (1987) *Processus et Mesure de l'Erosion*, Paris: CNRS, 438–454.

Harvey, A.M. (1987b) Sediment supply to upland streams: influence on channel adjustment. In Thorne, C.R., Bathurst, J.C. and Hey, R.D. (eds) (1987) *Sediment Transport in Gravel-bed Rivers*. Chichester: Wiley, 121–150.

Harvey, A.M., (1989). Reading the landscape: A. M.Harvey looks at the interaction between slope and channel form in the Howgill Fells. *Geographical Magazine: Analysis Supplement 19*, 6-7.

Harvey, A.M. (1991) The influence of sediment supply on channel morphology of upland streams: Howgill Fells, northwest England. *Earth Surface Processes and Landforms* **16**, 675–684.

Harvey, A.M. (1992) Process interactions and the development of hillslope gully systems: Howgill Fells, northwest England. *Geomorphology* **5**, 323–344.

Harvey, A.M. (1994) Influence of slope–stream coupling on process interactions on eroding gully slopes: Howgill Fells, northwest England. In Kirkby, M.J. (ed.) (1994) *Process Models and Theoretical Geomorphology*, Chichester: Wiley, 247–270.

Harvey, A.M. (1996) Holocene hillslope gully systems in the Howgill Fells, Cumbria. In Anderson, M.G. and Brooks, S.M. (eds) (1996) *Advances in Hillslope Processes*, Volume 2, Chichester: Wiley, 731–752.

Harvey, A.M. (1997) Coupling between hillslope gully systems and stream channels in the Howgill Fells, northwest England: temporal implications. *Geomorphologie, Relief, Processus, Environment* **1**, 3–20.

Harvey, A.M. (2000) Coupling within fluvial systems: spatial and temporal implications. *Journal of China University of Geosciences* **11**, 9–27.

Harvey, A.M. (2001) Coupling between hillslopes and channels in upland fluvial systems: implications for landscape sensitivity, illustrated from the Howgill Fells, northwest England. *Catena* **42**, 225–250.

Harvey, A.M. (2002) Effective timescales of coupling within fluvial systems. *Geomorphology* **44**, 175–201.

Harvey, A.M. (2007) Differential recovery from the effects of a 100-year storm: significance of long-term hillslope–channel coupling: Howgill Fells, northwest England. *Geomorphology* **84**, 192–208.

Harvey, A.M. (2012) The coupling status of alluvial fans and debris cones: a review and synthesis. (State of Science paper) *Earth Surface Processes and Landforms* **37**, 64–76.

Harvey, A.M. and Calvo-Cases, A. (1991) Process interactions and rill development on badland and gully slopes. *Zeitschrift fur Geomorphologie* Supplement **83**, 175–194.

Harvey, A.M. and Chiverrell, R.C. (2004) Carlingill, Howgill Fells. In Chiverrell, R.C., Plater, A.J. and Thomas, G.S.G. (eds) (2004) *Quaternary of the Isle of Man and Northwest of England – Field Guide*. Quaternary Research Association, London, 177–193.

Harvey, A.M., Hitchcock, D.H. and Hughes, D.J. (1979) Event frequency and morphological adjustment of fluvial systems in upland Britain. In Rhodes, D.D. and Williams, G.P.(eds) (1979) *Adjustments of the Fluvial System*. Dubuque, Iowa: Kendall/Hunt, 139–167.

Harvey, A.M., Alexander, R.W. and James, P.A. (1984) Lichens, soil development and the age of Holocene valley-floor landforms, Howgill Fells, Cumbria. *Geografiska Annaler* **66A**, 353–366.

Harvey, A.M., Oldfield, F., Baron, A.F. and Pearson, G. (1981) Dating of post-glacial landforms in the central Howgills. *Earth Surface Processes and Landforms* **6**, 401–412.

Jarman, D., Wilson, P. and Harrison, S. (2013) Are there any relict rock glaciers in the British mountains? *Journal of Quaternary Science* **28**, 131–143.

King, C.A.M. (1976) *Northern England: The Geomorphology of the British Isles*. London: Methuen, 211pp.

Lewin, J., Macklin, M.G. and Johnstone, E. (2005) Interpreting alluvial archives: sedimentological factors in the British Holocene fluvial record. *Quaternary Science Reviews* **24**, 1873–1889.

Livingstone, S.J., Evans, D.J.A. and Cofaigh, C.O. (2010) Re-advance of Scottish ice into the Solway Lowlands (Cumbria, UK) during the Main Late Devensian deglaciation. *Quaternary Science Reviews* **29**, 2544–2570.

Livingstone, S.J., Evans, D.J.A., Cofaigh, C.O., Davies, B.J., Merritt, J.W., Huddart, D., Mitchell, W.A., Roberts, D.H. and Yorke, L. (2012) Glaciodynamics of the central sector of the British–Irish Ice Sheet in Northern England. *Earth Science Reviews* **111**, 25–55.

Macklin, M.G. and Lewin, J. (2003) Holocene river alluviation in Britain. *Zeitschrift fur Geomorphologie*, Supplement **88**, 109–122.

Macklin, M.G., Passmore, D.G. and Rumsby, B.T. (1992) Climatic and cultural signals in Holocene alluvial sequences: the Tyne Basin, Northern England. In Needham, S. and Macklin, M.G. (eds) (1992) *Alluvial Archaeology in Britain*. Oxford: Oxbow Press, 123–139.

McConnel, R.B. (1939) The relict surfaces of the Howgill Fells. *Proceedings of the Yorkshire Geological Society* **24**, 152–164.

Miller, S.Y. (1991) *Soil chronosequences and fluvial landform development*. PhD Thesis, University of Liverpool.

Mitchell, A.F. (1991) Cautley Crags. In Mitchell, W.A. (ed.) (1991) *Western Pennines/Field Guide*, Quaternary Research Association, London, 94–98.

NERC (Natural Environmental Research Council) (1975) *Flood Studies Report* (5 volumes).

Oldfield, F. (1963) Pollen analysis and man's role in the ecological history of the southeast Lake District. *Geografiska Annaler* **45**, 23–40.

Pennington, W. (1991) Palaeolimnology of the English Lakes – some questions and answers over fifty years. *Hydrobiologia* **214**, 9–24.

Rumsby, B.T. and Macklin, M.G. (1994) Channel and floodplain response to recent abrupt climatic change: the Tyne basin, northern England. *Earth Surface Processes and Landforms* **19**, 499–515.

Schumm, S.A. (1956) Evolution of drainage systems and slopes in badlands at Perth Amboy, New Jersey. *Geological Society of America*, Bulletin **56**, 597–646.

Sissons, J.B. (1980) The Loch Lomond Advance in the Lake District, northern England. *Transactions of the Royal Society of Edinburgh, Earth Sciences* **71**, 13–27.

Stuiver, M. and Reimer, P.J. (1993) Extended database and revised CALIB radiocarbon calibration programme. *Radiocarbon* **35**, 215–230.

Stuiver, M., Reimer, P.J. and Reimer, R.W. (2005) CALIB 5.0 [WWW program and documentation]. http://calib.qub.ac.uk/calib/.

Tipping, R. and Halliday, S.P. (1994) The age of alluvial fan deposition at a site in the Southern Uplands of Scotland. *Earth Surface Processes and Landforms* **19**, 333–348.

Tipping, R., Millburn, P. and Halliday, S.P. (1999) Fluvial processes, land-use and climatic change in upper Annandale, southern Scotland. In Brown, A.G. and Quine, T. (eds) (1999) *Fluvial Processes and Environmental Change*. Chichester: Wiley, 311–328.

Turner, J. and Hodgson, J. (1979) Studies in the vegetational history of the northern Pennines: I. Variations in the composition of the early Flandrian forests. *Journal of Ecology* **67**, 629–646.

Turner, J. and Hodgson, J. (1983) Studies in the vegetational history of the northern Pennines: III. Variations in the composition of the mid-Flandrian forests. *Journal of Ecology* **71**, 95–118.

Turner, J. and Hodgson, J. (1991) Studies in the vegetational history of the northern Pennines: IV. Variations in the composition of the late-Flandrian forests and comparisons with those of the early and mid-Flandrian. *New Phytologist* **117**, 165–174.

Wainwright, A. (1972) *Walks in the Howgill Fells and adjoining fells*. (2003 Edition) London: Frances Lincoln, 174pp.

Wells, S.G. and Harvey, A.M. (1987) Sedimentologic and geomorphic variations in storm generated alluvial fans,

Howgill Fells, northwest England. *Geological Society of America Bulletin* **98**, 182–198.

Wilson, P., Lord, T. and Rodés, A. (2013) Deglaciation of the eastern Cumbria glaciokarst, northwest England, as determined by cosmogenic nuclide ([10]Be) surface exposure dating, and the pattern and significance of subsequent environmental changes. *Cave and Karst Science* **40**, 22–27.

Winchester, A.J.L. (1987) *Landscape and Society in Medieval Cumbria*. Edinburgh: J. Donald Ltd., 177pp.

Winchester, A.J.L. (2000) *The Harvest of the Hills: Rural Life in Northern England and the Scottish Borders 1400–1700*. Edinburgh: Edinburgh University Press, 194pp.

INDEX